THE ART AND SCIENCE OF
WIND POWER

Wind Turbine Technology and Design

David A. Rivkin, PhD

Kathleen Toomey, BS

Laurel Silk, MEd, BA

JONES & BARTLETT
LEARNING

World Headquarters
Jones & Bartlett Learning
5 Wall Street
Burlington, MA 01803
978-443-5000
info@jblearning.com
www.jblearning.com

Jones & Bartlett Learning books and products are available through most bookstores and online booksellers. To contact Jones & Bartlett Learning directly, call 800-832-0034, fax 978-443-8000, or visit our website, www.jblearning.com.

Wind Turbine Technology and Design is an independent publication and has not been authorized, sponsored, or otherwise approved by the owners of the trademarks or service marks referenced in this product. Some images in this book feature models. These models do not necessarily endorse, represent, or participate in the activities represented in the images.

This publication is designed to provide accurate and authoritative information in regard to the subject matter covered. It is sold with the understanding that the publisher is not engaged in rendering legal, accounting, or other professional service. If legal advice or other expert assistance is required, the service of a competent professional person should be sought.

Production Credits

Chief Executive Officer: Ty Field
President: James Homer
SVP, Editor-in-Chief: Michael Johnson
SVP, Chief Technology Officer: Dean Fossella
SVP, Chief Marketing Officer: Alison M. Pendergast
VP, Manufacturing and Inventory Control:
   Therese Connell
SVP, Curriculum Solutions: Christopher Will
Director of Sales, Curriculum Solutions: Randi Roger
Editorial Management: High Stakes Writing, LLC,
   Editor and Publisher: Lawrence J. Goodrich
Managing Editor, HSW: Ruth Walker
Copy Editor, HSW: Sarah Call

Senior Editorial Assistant: Rainna Erikson
Production Manager: Susan Schultz
Production Editor: Keith Henry
Production Assistant: Kristen Rogers
Senior Marketing Manager: Andrea DeFronzo
Manufacturing and Inventory Control Supervisor:
   Amy Bacus
Composition: Cenveo Publisher Services
Cover Design: Kristin E. Parker
Rights & Photo Research Associate: Lian Bruno
Cover Image: © Rui Vale de Sousa/ShutterStock, Inc.
Printing and Binding: Edwards Brothers Malloy
Cover Printing: Edwards Brothers Malloy

ISBN: 978-1-4496-2457-6

Library of Congress Cataloging-in-Publication Data
Unavailable at time of publication.

6048

Printed in the United States of America
16 15 14 13 12   10 9 8 7 6 5 4 3 2 1

Brief Contents

Contents

Preface

THE WIND ENERGY INDUSTRY is at the forefront of the world's shift away from reliance on fossil fuels. In just a few short decades wind energy has evolved dramatically. Technological advances now make wind energy a cost-effective solution for the world's ever-growing energy needs. The United States is now one of the world's leaders in overall wind power capacity.

As the wind energy industry continues to expand in the United States and around the globe, it will provide many opportunities for workers in search of new careers. These careers extend beyond the wind farm and include the efforts of employees who work in manufacturing plants, offices, and construction, as well as operation and maintenance. According to estimates from the American Wind Energy Association, approximately 85,000 Americans are currently employed in the wind-power industry. Despite the growing demand for skilled workers, there remains a lack of serious educational resources to meet the market's demand.

The *Art and Science of Wind Power* series was developed to fill this education gap. Each book in the series examines performance challenges using a systems perspective. Readers do not learn design and installation steps in a vacuum—instead they examine interrelationships and discover new ways to improve their own systems and positively contribute to the industry.

This series was developed for both the novice and expert. The texts take the learner from an overview of wind energy, through design and installation steps and considerations, to the design and installation of commercial wind systems.

Wind Turbine Technology and Design gives readers a basis for exploring the relationship between engineering design and wind-turbine economics. Readers gain an overview of large wind-turbine technologies and the design of rotors, drive trains, electrical systems, and towers. Topics covered include wind-turbine location, systems design, and integration as they affect wind energy's environmental impact.

About the Authors

Prof. David A. Rivkin, PhD, is managing director and dean of the College of Science and Technology at the Sustainable Methods Institute (SMI), an online university and innovation center.

He is also the chairman of the Department of Nanosciences in Renewable Energy at Chiist University, Atlanta, Georgia; and dean of education and research at the Israel Sustainability Institute. He is also chief scientist and director at the Adamah Group and its wind power division, in Israel.

Professor Rivkin was the founder, associate professor, and chairman of the Green Technologies Department at Ohalo College of Katzrin; associate professor at National University, based in San Diego, Calif.; International Technological University, in San Jose, Calif.; and the Graduate School of Science and Technology at the Technical University of Munich, in Germany.

He holds dual bachelor's degrees in chemistry and nuclear engineering from the University of California at Berkeley. Professor Rivkin pursued postgraduate studies toward a doctorate in biophysics at the University of California and later completed a PhD in business sciences at the European School of Business London at Regent's College, with a focus on small business sustainability. He has over 25 years of professional experience in both industry and academia.

He has taught at internationally renowned colleges and universities in Europe, India, China, the United States, and Israel. In 2010 Professor Rivkin was nominated to be a Fulbright scholar. He is also an Institute of Electrical and Electronics Engineers (IEEE) senior member and distinguished lecturer, a principal adviser in clean technologies to the National Science Foundation, a certified program manager, a certified corporate sustainability expert, and a certified green energy professional.

He is the winner of numerous technical and managerial awards and has been recognized for his outstanding contributions by governments as well as the United Nations. A serial entrepreneur, with roles in several successful ventures, including as founder and chief of technology for SciEssence International, Professor Rivkin has a multidisciplinary background, from biosciences to nanotechnology from health to energy, that gives him broad expertise in sustainability.

Kathleen Toomey is an experienced technical writer specializing in sustainable and cutting-edge technology. Following her initial career path in industrial construction, Toomey leveraged her engineering experience into a successful writing career. Currently, she focuses her skills and intentions on the environmental and renewable energy industries. Toomey has a BS in agricultural engineering from the University of Idaho.

Laurel Silk has managed e-learning initiatives at three leading universities in Arizona including Arizona State University, University of Phoenix, and Grand Canyon University. Among the highlights of her career, she created a virtual doctoral library for research students and designed and implemented a web-based doctoral studies program in administration. Ms. Silk is a former classroom instructor, in which capacity she designed and taught courses in freshman English, world literature, and critical thinking. As vice president of SilkWeb, she has created undergraduate and graduate online courses in higher education, renewable energy, business, and nursing. Ms. Silk holds a Masters degree in education with a focus on adult learning and instructional design technologies from University of Phoenix and has a Bachelor of Arts in English from Arizona State University.

Wind Technology and Design Overview

LARGE WIND TURBINE TECHNOLOGY has been one of the fastest-growing alternative energy resources since the mid-1980s. In earlier times, wind towers were considered a novelty with no real impact on the production of electricity. Today, they play an active role in the future of electric power generation.

Commercial wind farms are sprouting up across the country, supplying electricity to existing power grids. Computer-assisted design and controls, as well as advanced construction materials, make the large wind turbine industry more viable than ever before. There is no conclusive design for future wind turbines as they reach megawatt capacities. Weight, manufacturing costs, and economic feasibility are the three main factors weighed against future design advances.

Chapter Topics

This chapter covers the following topics and concepts:

- Early developers of wind energy and their contributions
- The reemergence of wind energy technology
- Large wind turbine design decisions
- Design vs. cost of commercial wind turbine development

Chapter Goals

When you complete this chapter, you will be able to:

- Relate how today's commercial wind turbine designs evolved
- Understand terminology related to large wind turbine development
- Discuss the design considerations of commercial wind turbines
- Understand the influence of costs on wind energy design and production

Pioneers of Wind Generation of Electricity

The concept of harnessing the Earth's wind and converting it into electricity is not new. In the late 1800s traditional windmill structures were the norm. They furnished electrical **power**—energy that is capable of or available for doing work—effectively, but they lacked efficiency and abundance of capacity. Charles F. Brush and Poul la Cour engineered technological insights that catapulted the wind industry into new ways of thinking. Much of their inventive technology is still in use today.

In the 1930s wind-generated electricity stopped being a novelty. President Franklin Delano Roosevelt signed the Rural Electrification Administration (REA) into law, bringing electricity to rural communities for the first time. The Carter administration helped put into place the start of the energy movement when the oil embargoes of the 1970s raised oil prices to such extremes. The energy movement moved forward at a slow pace until deregulation forced utility companies to buy alternative generated electricity using existing power grids.

Charles F. Brush

Charles F. Brush (1848–1929) was an entrepreneur and inventor from Cleveland, Ohio. He built what is considered to be the world's first fully automated wind-powered electric generator in 1887 **FIGURE 1-1**. It looked like a modified windmill and produced 12 kilowatts at its peak production. Overall the 40-ton steel tower stood 60 feet in height. The windmill itself encompassed a 56-foot diameter with 144 blades and a sail surface of 1,800 square feet. Inside the tower **shaft**, where the power is transferred, were pulleys, belts, and the dynamo (generator).

His automated control principles were used extensively throughout the industry until the 1980s when computer-process control became available. Brush is considered one of the main founders of the North American electrical industry. Commercializing his endeavors, he established the Brush Electric Company, which was sold and then merged in 1892 with Edison General Electric under the name General Electric (GE).

FIGURE 1-1 Brush's fully automated wind-powered electric generator.

Poul la Cour

Poul la Cour (1846-1908) was a scientist, inventor, and educator from Askov, Denmark. He laid the foundation for modern wind turbine technology with his testing of aerodynamic theories on small windmill models in a wind tunnel. These testing results produced theories that are still recognized and in use today. He discovered the key to producing the maximum amount of energy was not in the traditional windmill style.

Poul la Cour concluded that a fast rotating windmill with few rotor blades was more efficient **FIGURE 1-2** . He went on to say, in order to produce the maximum amount of energy, the number of windmill wings (blades) should be small, their bevel small, and the speed of rotation fast. This concept tripled the 12-kilowatt output of the heavy, slow-moving Brush-style turbine to a 35-kilowatt output. These principles, along with modern materials and refined design concept, are continuing to multiply output today.

FIGURE 1-2 Two of Paul la Cour's test windmills in 1897.

US Rural Electrification Administration (REA)

Up until 1935, nine out of ten rural homes, farms, and sparsely populated communities were without electricity. In 1935, President Roosevelt signed Executive Order No. 7037, establishing the Rural Electrification Administration (REA). Through government-funded work and loan programs, electricity became a viable and commercial commodity available to everyone.

The goal of the government program was to provide rural dwellings with the same electrical advantages as larger industrial cities. The program promoted the creation of new public utilities with a massive growing need to generate electricity. Power grids fingered out from every Roosevelt New Deal dam, and coal and oil-fired generating plants spider-webbed our country from coast to coast.

Smith-Putnam

In the early 1940s, the S. Morgan Smith Company, a water turbine manufacturer, supplied financial backing to US engineer Palmer Cosslett Putnam. This backing allocated the support needed to design and build the first commercial wind turbine **FIGURE 1-3**. It is believed this was the first attempt by any manufacturer

FIGURE 1-3 Smith-Putnam wind turbine installed at Grandpa's Knob near Rutland, Vt., in the early 1940s.
Courtesy of NASA

to generate alternating current by means of the wind for interconnection with a distribution system. News media referred to it as the "Smith-Putnam First Large Wind Generator." It comprised a large steel structure of:

- 100-foot tower with 75-ton rotating unit
- Two stainless steel vanes, the size and shape of a bomber's wings
- Capability of generating 1,340 horsepower or 1,000 kW (enough electricity to light 2,000 homes)

It was not without problems. Sudden gales would raise the turbine's output in three seconds from 1,000 to 3,000 kW, overloading an unbraked generator. Consequently, the tower operated for only four years, supplying electricity to the Central Vermont Public Service Company, but set a new benchmark for other manufacturers to strive for in their design.

Ulrich Hütter

Ulrich Hütter pioneered the German windmill industry in the 1950s. He developed a series of advanced horizontal axis wind turbines of intermediate size that used fiberglass airfoil-type blades with variable pitch. This not only allowed for lighter weight, but also higher efficiencies. Such a design approach sought to reduce bearing and structural failure by shedding the aerodynamic **loads** rather than withstanding them as previously designed by other manufacturers. Loads absorb forces such as torque, thrust, moment, vibration, and resonant frequencies.

Hütter also allowed the **rotor**, which is the hub and blade assembly, to teeter in response to wind gust and vertical **wind shear**, which causes a dramatic shift in wind speed and direction. The hub is the center of a wind turbine rotor. The use of a flexible teetering **hub,** rather than a rigid hub, was a major advantage of the European manufacturers. It allowed them to dominate the global wind turbine market until adaptation by other countries' manufacturers. Hütter's variable pitch and a teetering hub are an important integral design function used extensively in production of commercial wind turbine design today.

The Reemergence of Wind Power Technology

Currently 85 percent of the electricity supplied to homes, factories, and commercial buildings in the United States is generated by the burning of fossil fuels in power plants. Cheap oil from abroad and the abundance of US fossil fuels—wood, coal, natural gas, and oil—held back the use of alternative energy throughout most of the 1900s.

A growing need for more electricity pushed commercial and utility power organizations to seek the most reliable, most easily sourced, cheapest fuels available; however, other factors started to surface. Environmental issues of clean air, clear waterways, industrial emissions, and carbon dioxide from auto exhaust began hitting the newspaper headlines. The United States was a nation awakening to its bad environmental habits.

Beginning of the Environmental Movement

Concern about nuclear energy and its radioactive waste contributed to the stirrings of the environmental movement. The use of nuclear energy was not fully understood by many people and it did not go over well in the public press. Two major disasters—at Chernobyl, in Ukraine, and Three Mile Island, in Pennsylvania—further heightened fears that radioactive nuclear waste could possibly mean the end of civilization.

The nuclear waste issue of safe containment served to galvanize individuals into action. This movement started a whole new generation of homemade backyard alternative energy machines, including wind turbines. New terminology, such as "renewable resource," "sustainable energy," and "stewardship of the land," was being used in the press, in schools, in corporate reports, and even in contracts.

The dangers of nuclear energy production have never been fully understood. This has helped feed fears and made it controversial. Proponents call nuclear energy the most viable resource currently available for meeting the world's growing energy needs. Nuclear energy opponents fear the byproduct of production. They contend the world could be in grave danger of nuclear waste leakage into the environment if containment were insufficient.

Nuclear energy fired up the environmental movement, which in turn led to increased interest in more earth-friendly alternatives. More emphasis is being placed on the cleanup of nuclear waste than on the enhancing and building of more plants. Whatever its future, nuclear energy did instigate the beginning of an environmental movement that climbs in popularity every year.

The 1970s Oil Crisis

The 1973 oil crisis began when members of the Organization of Arab Petroleum Exporting Countries or OAPEC (consisting of the Arab members of OPEC, plus Egypt, Syria, and Tunisia) proclaimed an oil embargo. Merchant ships were cut off from taking on oil from these countries destined for Western industrialized nations. The United States experienced long lines at the gas pumps, escalating fuel prices within months, and homes without heat.

In April 18, 1977, President Jimmy Carter gave a televised speech on the need to balance our demand for energy with our rapidly shrinking resources. Two days later, as promised in his speech, he presented his comprehensive energy program to the US Congress. This energy proposal established a new Department of Energy and the deregulation of energy.

The Carter administration made the public more aware of our energy needs and lack of exploration for alternative energy sources. Government support for the development of megawatt wind turbines was given in the anticipation they would provide a renewable energy technology in units that yield large enough quantity to be useful to public utilities. This was the tipping point for wind energy and the development of large turbines.

The California Wind Rush

The National Energy Act of 1978 was a legislative response by the US Congress to the 1973 energy crisis. Included in this act was the Public Utility Regulatory Policies Act (PURPA). The law was intended to promote a greater use of renewable energy.

The PURPA law mandated electric utilities buy power from alternative resource producers. It succeeded in creating a new market for non-utility electric power producers such as wind farms and large solar installations. This pivotal legislation enabled renewable energy providers to gain a toehold in the commercial market.

Public and private utilities were now deregulated to allow integration of alternative energy into the established grid system (highline wires that deliver electricity from hydro dams and fossil fuel power plants). These are the high steel crossbar towers that can be seen along many highways. They traverse the countryside connecting to local substations in both rural locations and cities. Once tied into this **grid**, alternative power could be transported and sold commercially alongside the traditional power of dams and power plants.

All this turned "alternative power" into a significant source for commercial energy companies. Connecting to these high lines has been the biggest major factor in the development of large wind turbine technology. These systems now have somewhere to sell the power being generated on a commercial scale, whereas before they were limited to individual investors and their own private sites.

The State of California took the biggest advantage of the new deregulation laws by giving tax breaks to investors, over and above the tax incentives offered by the federal government. The sudden interest and economic potential of wind farms in the 1970s in California is sometimes referred to as the California wind rush because of the similarity it shares with the California gold rush of the nineteenth century. The fervor in which wind farms were constructed in California in the 1970s shared the same excitement and hope for good fortune.

In 1986 and 1987, the government stopped providing tax credits for private investors. In addition, when the cost of oil dropped significantly, so did the costs to run conventional power plants. These two factors made wind farm technology much less attractive. (A **wind farm** consists of any number of wind towers within close proximity of one another connecting to a common electrical transfer or substation.) Wind turbine manufacturers continued to build, however, because the PURPA deregulation was enough incentive to continue developing major commercial wind facilities.

Today's Commercial Wind Turbines

In the past, wind turbines oftentimes were looked upon as a novelty, always in the experimental stages, and open to reinvention. The 1980s deregulation enabled

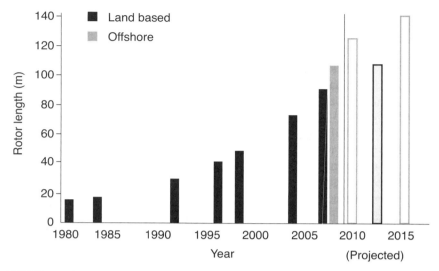

FIGURE 1-4 Increase in large wind turbine kilowatt output capabilities from 1980 after the advent of deregulation and computer aided design.

Data from Plantier, K., & Smith, K. M. (2009). *Electromechanical Principles of Wind Turbines For Wind Energy Technicians, 1st ed.* Waco, TX: TSTC Publishing.

the wind industry to design bigger and better turbines on a large scale. They now had a commercial marketplace to sell into. In the past, wind turbines were measured in terms of kilowatts, but the new generation of wind turbines is referred to in megawatts (MW).

Machine costs declined, because of lighter-weight materials of construction, improved aerodynamic design, and increased tower height. This relates to a significant reduction in unit costs of electricity production. Wind turbine-generated electricity is a rapidly growing resource. By the end of 1980, it encompassed 1 percent of worldwide energy production, increasing 500 percent globally by 2000–2007 **FIGURE 1-4**.

Until most recent years, each wind tower assembly was constructed individually. Today commercial wind turbines are built on assembly lines. This has helped greatly to bring down not only the cost of development but the cost of production as well.

Design Decisions

Wind towers may be a complex dynamic system but the four main components are relatively straightforward: blades, rotor hub, generator, and tower **FIGURE 1-5**. Rotating blades capture and harness the free flow of the wind while the rotor hub converts this natural **energy** into mechanical motion (power). The generator converts the mechanical power into electricity, and a tower is used for support and

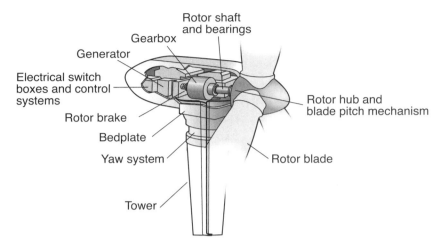

FIGURE 1-5 Main components of the most common wind tower configuration.

Adapted from Manwell, J. F., McGowan, J. G., and Rogers, A. L. (2009). *Wind Energy Explained: Theory, Design, and Application.* West Sussex, England: John Wiley & Sons Ltd.

conveyance of that electricity to the ground for regulation and transmission.

It is when you add weight, stress, and load forces, along with economical feasibility, that design decisions become critical. Blade selection is the first and foremost issue. The number of blades, length of the blade, style, and design of the blade will greatly influence and drive the selection and cost of other components associated with the tower turbine.

> **NOTE**
> Wind turbine configurations are many and vastly varied. As an example, some turbines have a fixed speed and no gearbox is required. Variances in system configurations are numerous from manufacturer to manufacturer, and over the scope of turbine size.

Number of Blades

Wind towers can have as many blades as their rotor assembly will hold. However, as Poul la Cour discovered in the late 1800s, the fewer blades used, the more efficient the turbine. A two-blade or three-blade tower is the most prevalent configuration for large wind turbines today; three blades are more conventional for commercial applications **FIGURE 1-6** .

A two-blade rotor assembly will have less weight than a three blade, but they require a higher blade tip speed for optimum efficiency resulting in higher noise emissions. They also do not produce a symmetric load (equal force in all directions) as nicely as a three-blade system. A two-blade tower experiences wind shear when the two blades are directly inline with the vertical position of the tower. At that particular point of position, the wind shear they produce causes a bending

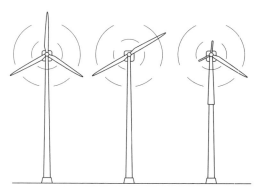

FIGURE 1-6 The three most prominent rotor blade configurations in production today: three-, two-, and one-blade turbines.

Adapted from Plantier, K., & Smith, K. M. (2009). *Electromechanical Principles of Wind Turbines For Wind Energy Technicians, 1st ed.* Waco, TX: TSTC Publishing.

effect on the turbine drive shaft inside the **nacelle**, which houses the components of the wind turbine.

A three-blade configuration decreases the effects of shear because its blades are offset. Only one blade at a time is inline with the tower while the other two both counterbalance at a 120-degree angle for a more dynamically balanced rotor load.

One-blade turbines offer no weight reduction benefit due to the counter-balance required to offset the single blade weight. This imbalance makes the one blade much more complex even though a reduction in the number of blades results in increased rotor speed and lower drive train loads. A one-blade turbine has the same wind shear effect as a two-blade configuration and will produce higher noise emissions resulting in considerable concern for populated areas.

Blade Radius

Quite often blade radius is referred to as simply the rotor blade length. This is too simplistic and inaccurate for the needs of either an engineer or a designer, however. A designer measures the blade radius from the center of the rotor hub to a specified length on the blade for calculation of that section's airfoil dynamics **FIGURE 1-7** . This is important to note, as all blade design calculations refer to this distance and not the overall distance of a blade.

The shape of a rotor blade is not symmetrical. Calculations are done for a number of blade radius set points, each having its own design parameters. Blade radius, chord length, and pitch (blade angle) are the three dimensions needed to calculate the shape (aerodynamics) of a rotor blade.

NOTE

The longer the overall blade length, the more wind energy it will capture.

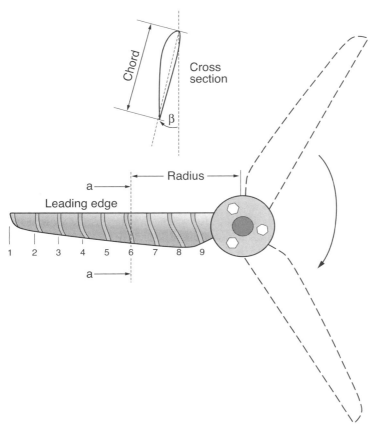

FIGURE 1-7 Blade radius dimension for a particular cross section (a-a) of a rotor blade.
Adapted from Warlock Engineering

Blade Twist Distribution

The difference in pitch (blade angle) to line of blade travel at different blade radius points is called blade twist distribution. A twist on the **trailing edge** of a blade, which faces away from the direction of rotation, will help distribute and maximize wind flow along the full length of the blade. The **leading edge** of a blade, which faces toward the direction of rotation, does not receive as much twist and in some cases none at all **FIGURE 1-8**.

Generally, blade twist allows more pitch at the blade root for easier startup, and less pitch at the tip for better high-speed performance. Although it is much simpler to produce a rotor blade without a twist, experts agree that a twist-free blade will contribute to a loss in power.

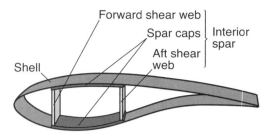

FIGURE 1-8 Rotor blade terminology.

Adapted from Manwell, J. F., McGowan, J. G., and Rogers, A. L. (2009). *Wind Energy Explained: Theory, Design, and Application.* West Sussex, England: John Wiley & Sons Ltd.

Airfoils

When assessing rotor blade airfoils, the variable that carries the most importance is called the *lift-to-drag ratio*, also written as L/D **FIGURE 1-9**. Remember that high lift and low drag is optimal. When the L/D ratio decreases, so does the power coefficient. Interesting to note is the difference between low-speed and high-speed rotors. The L/D ratio is of little significance in rotors of lower speeds. However, with high-speed rotors the L/D ratio makes a substantial difference in how much power can be generated.

> **NOTE**
>
> In the wind turbine design industry, airfoil is often times referred to as aerofoil.

Revolutions per Minute

A turbine's **revolutions per minute (rpm)** capacity refers to how fast an individual blade rotates through the swept area (the area within the circumference scribed by the blade tip), which in turn transfers that rotating motion to the axis of the rotor assembly in the hub of the turbine. It stands to reason, the faster the speed of rotation (i.e., rpm) around the horizontal axis, the greater the amount of wind gathering potential.

Regardless of the availability of wind, it is the design of the blade and rotor assembly that will dictate a tower's overall rpm rating. If an rpm is too slow, it will stall. A high rpm results in mechanical stress; wind gusts are the most prevalent of

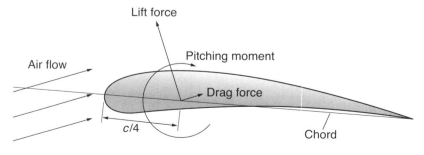

FIGURE 1-9 Airfoil showing lift-to-drag forces.

issues. There is optimum speed to every turbine system depending on the rotor-to-generator ratio.

Variable RPM, Variable Pitch

Connecting a wind tower directly to a grid enables it to have a fixed speed whereby the rotor and generator must produce a consistent 60 Hz at all times. Because wind does not flow at a constant rate, and wind gusts most assuredly play havoc with rotor speed and turbine drive trains, variable rpm and pitch are used. Almost all turbines today have variable rpm and variable pitch to enhance and take advantage of wind velocity fluctuations for a greater bandwidth of power output.

Variable rpm allows for more consistent run time with higher efficiency in capturing the wind and fewer maintenance issues. However, power generated at variable speeds will need to be regulated before it is transferred to the power grid. There is a cost tradeoff between fixed and variable speed, since additional equipment is needed to boast and regulate the generated power. Additional equipment means additional weight at the top of the tower, or **towerhead**, which can affect the cost of construction and cause maintenance concerns. The **tower** is the structure used to support the towerhead.

A *variable pitch* (change of angle-of-attack) can be either active or passive. Active pitch control requires instrumentation in and on the tower nacelle, and mechanical equipment inside the rotor hub. By taking wind readings, the active pitch can adjust the pitch (blade angle) to take into maximum account of the wind for optimum performance. Passive pitch control relies on the blade itself adjusting to the wind. This method of control is best used on smaller commercial machines.

Cost, Noise, Vibrations, and Fatigue Considerations

The cost of construction and maintenance of a wind tower figures heavily in whether a tower successfully produces economical power. The power production of a wind farm must be competitive with the already established commercial market. It is a juggling act to access every component of a wind tower for cost and compare it against numerous component choices, each having its own cost-to-design advantages.

As an example, one machine design may be lightweight, based upon a two-blade rotor concept. But a two-blade system involves increased noise emissions, vibration, and fatigue constraints that must be analyzed and factored into the equation. If the tower is to be located near a residential area, noise levels become more of a concern.

Noise levels from a wind tower are more in a low-frequency range resulting from the displacement of air by the blades and turbulence at the blade surface. Any scratch, dirt, insect, or irregularity on the blade surface at high speed promotes a noticeable audible sound. In addition, a slit or cut parallel to the trailing edge will emit a distinguishable constant pitch of high frequency. Noise is also produced as a rotor blade passes in front of the tower. The sudden change in wind direction, such as the wind shear that a blade encounters when it passes the tower, will set up a turbulent condition resulting in another audible sound source.

Improved designs have reduced the vibration sound originating from turbines themselves. Such noise has now become irrelevant to neighbors. However, from a design standpoint, any vibration is an issue, because vibration is indicative of possible wear and fatigue on the machine and its components parts.

The two types of fatiguing loads are fluctuating and cyclic. Fluctuating loads originate from variable air turbulence and wind gust. Cyclic loads are deterministic. They are generated via the self weight of the tower rotating components and the cyclic bending forces of the blades. Again, as the blades rotate past the tower, a wind shear effect places great burden on the turbine mechanical systems. Designers must design for both extreme and normal loading conditions, and normal internal operating conditions—all in conjunction with efforts to maximize output while minimizing load forces.

The load forces across the swept area of the blade are considerable. They vary from the tip of the blade to its root at the hub. Combine this with weight and it adds up to a substantial increase in fatigue. A mere lengthening of the rotor blades radiates fatigue—not only the length of the blade itself, but all the way back to the main structure of the rotor hub, to the drive shaft and gearbox, and ultimately even back to the support tower.

Materials of Construction

Materials of construction can vary greatly from manufacturer to manufacturer as they strive to lower weight-bearing loads. Weight is first and foremost the most sought-after factor behind material selection. It is the biggest deterrent to efficiency and starts with the rotor blade design. The larger the swept area, the more power a turbine is capable of extracting from the wind. However, each extra measure of blade length requires extra strength and adds further to its structural weight.

Tensile strength—the strength of a material expressed as the greatest lengthwise stress it can bear without tearing apart—and density are the two main parameters influencing the material selection of rotor blades. The weight of a blade is inversely proportional to its specific strength. Today's commercial rotor blades most typically use some derivative of spun plastic fiberglass. Glass-fiber reinforced polyester (GRP) or glass-fiber reinforced epoxy (GRE) are two of the most common.

These fiberglass products are composite material made of plastic molds reinforced with fine fibers made of glass. GRP has a good weight-to-strength ratio and rust resistance, and can be molded in a variety of ways. GRE is gaining popularity as the use of polyester becomes less attractive because of environmental restrictions. GRE also has the advantage of being used in the form of pre-impregnated laminate, enabling improved quality control.

Development of Commercial Wind Turbines

Even though the wind blows around us everywhere in abundance and costs nothing, do not assume electricity supplied by harnessing the wind's energy is cheap. The cost to research, develop, capture, and harness the wind greatly offsets this free availability.

Only in recent years has wind energy been commercially exploited for public utility production. This is due in part to the advent of deregulation but more important, to the development of large wind turbine technology. Today's wind turbines are now on a megawatt scale, and they are being used not just singly, but in wind farms. Large commercial wind turbines and their respective production are in many instances now competing on their own, supplying power into existing power grids across the nation.

The cost of developing a new technology is high, but it's important to understand the cost saving improvements of **value engineering**. In this phase of product development, all aspects of the design are scrutinized for ways and means to deliver product that is lighter and less prone to stress fatigue. Weights are toyed with and designs are adjusted until the right balance of weight to efficiency is obtained. Many steps are taken with this phase to produce a viable and profitable wind collection system.

Rating Large Wind Turbines

Wind holds within its raw state the potential energy needed for electricity. Only when it is changed into a measurable state can it be rated. That state is electrical power. The rotor blades of a wind turbine collect the erratic energy of the wind and convert it into mechanical motion via a rotating drive shaft. The drive shaft warrants this mechanical power measureable via the rate of rpm. Next, the drive shaft gears engage to mechanically step up this calculable capacity of power and a generator transforms it into electrical power, measured by a unit known as the kilowatt.

This description greatly simplifies the many sub-systems, controls, instrumentation, and regulation of a wind turbine. However, now you have a reference in which to compare the capacity of one wind turbine to another. The power rating (watt) represents not only the power capability of the wind turbine, but its size. Typically the higher the number of watts, the longer the rotor blades will be.

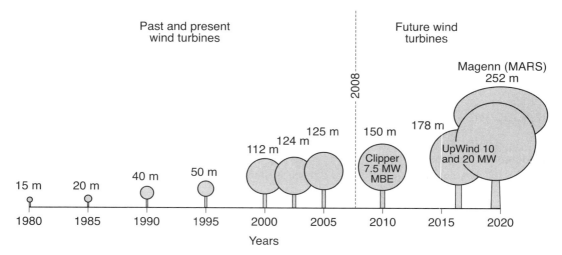

FIGURE 1–10 Size comparison of past and future wind turbines by swept area and MW rating (future).

Adapted from Plantier, K., & Smith, K. M. (2009). *Electromechanical Principles of Wind Turbines For Wind Energy Technicians, 1st ed.* Waco, TX: TSTC Publishing.

Wind turbine size, expressed as swept area, has increased steadily as turbines have developed. Kilowatt (kW) and megawatt (MW) are used to rate large wind turbines today FIGURE 1–10.

Size matters, but is bigger always better? Oftentimes the megawatt turbines are efficient only at higher wind velocities. The blades and gearing are heavier than in smaller turbines and not so easily moved in light-to-lower velocity winds. Many wind farms take advantage of lower wind velocity and collect from many towers to pass the cumulative energy along at greater savings. Also, with the advent of revolutionary materials and design, wind power manufacturers are showing greater interest in the smaller turbine with higher impact—power generation. Either scenario will still adhere to kW representing the final factor for rating the commercial wind turbine.

Cost of Research and Development

Research and development (R&D) is the most costly phase of getting a new design into production. It is the first in a sequence of three stages, whereby the initial concept and the exploring of all design avenues and testing begin. Next, a workable prototype is created for manufacturing to build. It's important to note that an R&D team should strive to create a prototype that is cost effective, thus assuring the manufacturer the new design will be competitive in the marketplace.

Research and development, however, does not stop with the prototype. The second stage closely tracks the first but places more emphasis on working designs with continued efficiency of the systems after the tower units are in service. Maintenance is a key concern at this juncture, as is safety and the monitoring of performance—not just for performance output but also for cost of operation, which relates to the profitability of the design. Once towers are in place, feedback from the field is taken into consideration. Testing and monitoring are taken under advisement with an aim for future breakthroughs.

Once the R&D groups have working, surviving designs, Stage 3 is reached. Here the emphasis on value engineering. The concern now is to reduce cost. The manufacturer wants to attract the commercial market at a competitive and economically attractive level. Value engineering is applied to reduce costs while still meeting the design criteria, and as a means of future design improvement. Value engineering also involves means and methods to cut costs further through plant tooling and production methods.

Cost of Marketing and Production

Cost of production is where the prototype goes into full plant production, and marketing ramps up. Tooling of the production plant, engineering optimization, and still more reduction in costs are the major concerns at this stage.

Development is still involved, as advances in technology are taken advantage of and operational experience is taken under advisement. Concepts have changed and technology has moved forward, so small improvements in design can still be actualized. The cost savings that can be realized at this point tend to be minimal, but they tend to continue throughout the course of production and they add up.

Also, as marketing and sales capture orders and receive commitments to buy, prices for materials will be renegotiated. The manufacturer can now take advantage of and realize quantity pricing. Another cost reduction to consider is plant tooling for assembly-line production versus the one-at-a-time assembly of the prototype.

Cost of Series-Produced Wind Turbines

Series-produced turbines wire their generator alternators in series. This is a three-phase alternator wiring configuration in which all three phases are connected to one another. Each coil of each phase is connected to the one next to it, and so on, increasing voltage but leaving amperage the same.

Series-produced horizontal axis wind turbines have been the leading technology responsible for higher power ratings (kW). These higher power ratings are the key factor allowing wind technology to be taken seriously in the utility-scale commercial energy market.

FUTURE TRENDS OF WIND POWER TECHNOLOGY

A strong future for wind power expansion seems certain. The Global Wind Energy Council projects production of global wind power to reach 332 GW by 2013, almost triple the current output. Larger and more sophisticated towers in the 10–20 MW range are projected for the future. But bigger isn't always better. Advances in control technology mean more energy can be generated from smaller machines.

One example of advanced technology is the ultrasonic or laser-type anemometer. This device can detect the shifting of sound or coherent light reflected from the air molecules 100 feet out in front of the tower rotor to calculate the speed and direction of the wind before it comes in contact with the rotor blades. Being aware of the forces about to impact a tower before they strike could mean a major shift in design thinking for future towers!

Most important, deepwater offshore developments using MW turbines are going forward at a rapid pace. Offshore high-volume wind with more consistency means bigger tower designs. Europe has already been harvesting offshore power. The United States is following suit with its first offshore development on the east coast of Cape Cod.

CHAPTER SUMMARY

Large wind towers look so streamlined and simple standing tall and sleek on the horizon. Some people dislike their presence while others view them as moving works of art. No matter your stance or opinion, large commercial wind towers have made big inroads and become a viable source of electrical energy. No longer a novelty or "alternative" source, wind farms are now part of the power grid infrastructure. Their design concepts and material selections drive the industry forward to the future of megawatt producing turbines.

KEY CONCEPTS AND TERMS

Energy	Shaft
Grid	Tensile strength
Hub	Tower
Leading edge	Towerhead
Load	Trailing edge
Nacelle	Turbine
Power	Value engineering
Revolutions per minute (rpm)	Wind farm
Rotor	Wind shear

CHAPTER ASSESSMENT: WIND TECHNOLOGY AND DESIGN OVERVIEW

1. Who was the first inventor to fully automate the wind powered turbine?
   - ❑ **A.** Poul la Cour
   - ❑ **B.** Charles F. Brush
   - ❑ **C.** Edison
   - ❑ **D.** Smith-Putnam

2. Which US President initiated the deregulation of energy?
   - ❑ **A.** Theodore Roosevelt
   - ❑ **B.** Ronald Regan
   - ❑ **C.** Jimmy Carter
   - ❑ **D.** John F. Kennedy

3. Five MW represents 1,000,000 watts of electrical power.
   - ❑ **A.** True
   - ❑ **B.** False

4. What holds the blades of a wind turbine in place?
   - ❏ **A.** Rotor
   - ❏ **B.** Hub
   - ❏ **C.** Shaft
   - ❏ **D.** Nacelle

5. What does kilowatt (kW) represent?
   - ❏ **A.** Unit of energy
   - ❏ **B.** Unit of power
   - ❏ **C.** Joule
   - ❏ **D.** Electrical charge

6. What does an anemometer read on the outside of the tower nacelle?
   - ❏ **A.** Temperature
   - ❏ **B.** Humidity
   - ❏ **C.** Wind speed
   - ❏ **D.** Wind direction

7. Value engineering is used to _____ after a prototype is ready for production.

8. The Public Utility Regularly Polices Act (PURPA) allowed for deregulation of the power grid.
   - ❏ **A.** True
   - ❏ **B.** False

9. The nacelle houses which of the following?
   - ❏ **A.** Hub
   - ❏ **B.** Anemometer
   - ❏ **C.** Generator
   - ❏ **D.** Rotor

10. The longer the overall blade length the more wind _____ it will capture.

Wind Turbine Technology and Design Concepts, Part 1

LARGE WIND TURBINES, developed for commercial use, comprise two main design concepts: mechanical and electrical. This chapter will cover mechanical concepts, including rotor assembly, power control systems, blade pitch control, rotor brake systems, mechanical drive train, and nacelle.

This chapter also discusses the advantages and relative design choices to be made for each mechanical concept and subordinate subsystems. Each design choice taken will reflect a parameter that in turn will affect all other downstream components, starting with where the wind is first encountered at the rotor blade on through the drive shaft, gearbox, and nacelle.

Chapter Topics

This chapter covers the following topics and concepts:

- The rotor assembly and its configuration variables
- The three types of power control
- Blade pitch control
- The rotor braking system
- The mechanical drive train and the nacelle

Chapter Goals

When you complete this chapter, you will be able to:

- Identify each commercial wind turbine's main mechanical systems
- Relate the mechanical system concepts to each other

Rotor Assembly

The rotor assembly comprises the rotor blade(s) and a rotor hub that the blade(s) tie into. Overall, the rotor assembly undertakes to capture as much of the kinetic energy from the wind as efficiently as possible with the least amount of stress on the rest of the tower structure and systems. The efficiency of a wind turbine is measured by the **coefficient of power (C_p)**, often referred to as power coefficient. The coefficient is a number that represents a quantity of comparison; for example, the power coefficient represents the ratio of electricity produced by a wind turbine (kW) to the amount of wind energy converted.

Albert Betz, a German physicist, stated that no wind turbine could convert 100 percent of the wind to power **FIGURE 2-1**. By use of a lengthy and entailed physics calculation he devised the theoretical maximum value, known as the Betz limit, to be 59 percent for a **horizontal axis** wind tower. This figure most often is stated as a power coefficient of 0.59. Most horizontal axis wind turbines operate between a 0.34 and 0.50 power coefficient.

NOTE

It does not matter which direction the rotor assembly turns (clockwise or counterclockwise) to create an aerodynamic design. Clockwise is the accepted manner of rotation, as it matches up with the downstream mechanics of a typical turbine generator.

The swept area of the rotor assembly plays an important role in the calculation of this coefficient. Larger swept areas typically produce a higher power coefficient. As wind passes through the rotor assembly swept area, a certain amount of wind is spilled or not picked up by the blades. This will figure into the C_p calculations along with other ratio factors such as **tip-speed ratio**, which is the ratio between the rotational speed of the tip of the rotor blade and the actual velocity of the wind, along with the lift-to-drag ratio, and wind speed.

Configuration Variables

Aerodynamic efficiency, component costs, and system reliability are the primary factors when considering the design of the rotor assembly of a wind turbine design. The common denominator for developing and maintaining the rotor assembly is weight. And there are other design configurations to consider as well:

- The number of blades
- Flexibility of blade materials
- Materials used in manufacturing the blades

Theoretically, the number of blades per rotor assembly can be as many as the hub will allow. However, design, testing, and proven field production find the two- and three-blade assembly the most feasible. One-blade towers have the advantage of less weight but their need to be counterbalanced is a disadvantage. A two-blade assembly also gives a weight reduction, but wind shear and cyclic loads are a deterrent. One- and two-blades offer an increase in rotor tip speed, but produce significantly higher noise emissions than a three-blade assembly. One-blade

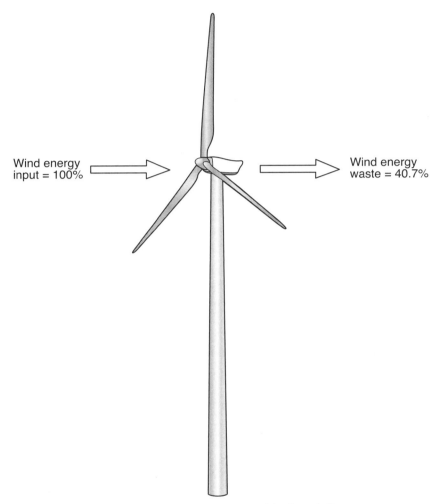

Wind energy
input = 100%

Wind energy
waste = 40.7%

FIGURE 2-1 No wind tower can capture 100 percent of the free wind stream energy.
Adapted from KidWind Project, Inc.

towers are still in the experimental stages, as they do offer cost benefits over and above their design constraints.

The three-blade assembly is more widely used on commercial wind towers and has a higher power coefficient, close to 0.50, than a two-blade assembly, which has a power coefficient of about 0.46. Three blades dynamically balance their load for less cyclic stress and fatigue on adjacent wind turbine components. Also, assemblies that have more blades usually have less of a spill factor as the wind passes through the blade swept area for a higher power coefficient. In addition, many people consider three-blade towers more aesthetically pleasing to look at, since they appear to be more balanced. The flexibility of each rotor blade is also a major consideration. A rigid blade has no capacity to shed gusts of wind or give to the sharp acceleration and fluctuations in the wind. Flexibility of the blade allows for adaptability to these changes. Blades with higher flexibility usually produce a more consistent speed of rotation.

By shedding the wind load of a blade, you also allow for less stress on the adjacent and downstream components of the wind turbine, and the tower structure itself. In design terms this all adds up to less stress, less maintenance, less weight, and less cost of construction. Sudden increases in blade load can be reflected throughout the tower system. These loads can travel along the blade to its base and back to the turbine main drive shaft and even to the tower itself. The more wind that can be shed off the blade edge, the more load forces it can take.

NOTE

It's important for the designer of a turbine to make sure the flexibility of a blade is not so great as to interfere with the tower during high wind gusts.

Blade flexibility allows for improved energy capture by making the blade able to stay more constant in its rotation and tip-speed ratio. By integrating flexibility into the blade, the designer will better ensure that dangerous tendencies of resonant frequencies, such as vibration and oscillation, are more likely to be diminished. Blade flexibility is achieved mainly by the use of composite materials in the blade's materials of construction.

Blade lengths in today's commercial towers are in the 100 to 175 foot range and involve various proven fiberglass composites. With prototype blades pushing to 200 feet, a wind manufacturer's never-ending quest is always a variation on the same theme—lighter, stronger blade materials.

A glass-fiber reinforced polymer composite material allows for high tensile strength and low density producing a blade that is easily bent or shaped. However, epoxy-based composites are piquing the interest of blade manufacturers because they offer a combination of environmental, production, and cost advantages. Carbon fiber composite materials are the most current design advantage. They are being identified as a cost-effective means for reducing weight and increasing stiffness while still allowing some flexibility. This is especially important as rotor blade lengths get longer.

Power Control

Depending on the rated power (kW) of a tower, wind turbines in general are designed to give maximum power output at wind speeds of about 33 mph. High wind acceleration and wind gusts can overload the towerhead causing immediate damage to the mechanical drive-train shaft and gearbox. Long-term consistent high wind speeds will cause fatigue of the electrical systems.

As these high wind conditions usually take place over a brief period of time, the power that could be generated has no significant overall value in terms of power output. Therefore, initiating a power control system to limit the rotor assembly speed will increase the overall mechanical power output to the turbine. This is the principal purpose of a power control system.

Designers working within the rated power of the towerhead parameters create a design so that the rotor assembly operates within a fixed bandwidth speed. The purpose of doing this is to avoid overtaxing or imposing too heavy a load on the various system components. There is a balancing relationship between power control and maximizing the efficiency of a tower **FIGURE 2-2**.

The three means of power control are:

- Pitch control
- Stall control
- Active yaw control

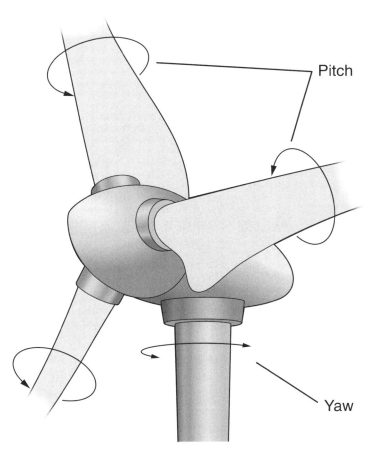

FIGURE 2-2 Wind tower showing axes of rotation for pitch and yaw.

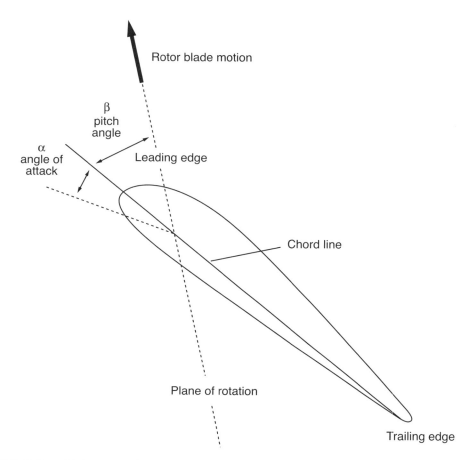

FIGURE 2-3 Rotor blade cross section showing pitch angle and angle of attack.
Adapted from Warlock Engineering

Pitch Control

Pitch control varies the orientation (angle of attack) of the rotor blades thereby alter-ing the aerodynamic lift-to-drag ratio of the blade. The rotor blade tie-in, which serves as a fastener to the hub is not fixed, but rotates around its longitudinal axis allowing for rotation of the blade and adjusting the angle of attack (pitch) **FIGURE 2-3**. This variable pitch affects the speed and efficiency of the rotor assembly. Wind turbine designers like a variable pitch in order to take full advantage of the wind forces and to make sure the turbines power output stays at its peak.

In large commercial wind turbines the blades are controlled via a variable pitch control system in the hub. This control is then carried out at the rotor blade by changing the lift-to-drag ratio dynamics of the blade. The variable pitch control is used to give an angle of attack out of alignment with the wind to produce more lift-to-drag ratio resulting in rotational movement of the rotor assembly.

The rotor blades change their pitch a fraction of a degree constantly while the rotor assembly is turning around the horizontal axis of the turbine. This allows for the power output to be more accurately controlled and optimized for a higher continuous output of power.

Stall Control

Stall control promotes a slowing down of the rotor assembly as a means of damping the lift-to-drag ratio when wind speeds become too excessive. It protects the towerhead from damage due to excessive wind gusts, or rapidly accelerating and fluctuating wind. It also is used for maintenance shut downs or when the system needs to be brought offline for whatever reason. When towers are first erected they are also kept in a stall position until ready to bring online.

Small- and medium-sized wind turbines with a fixed speed use a blade pitch angle that is fixed. The blades are bolted in place at the hub and not allowed to rotate around their longitudinal axis. They are designed to stall purposely at high wind speeds.

Stall control of a fixed blade happens automatically whenever an increase in the wind is above the blade's rated power. As the wind increases in speed, the wind flow separates from the blade surface and the lift forces of the blade decline while the drag forces increase giving an overall effect of stalling. This will keep the rotor assembly within a fixed speed and prevent the turbine generator from being overpowered.

Fixed blade design technology on the larger megawatt towers has not been successful due to lack of definitive system feedback and operational problems. Megawatt towers are more inclined to use an active stall control system with variable pitch rotor blades operating as part of the pitch control system.

A variable pitch rotor blade, when fully stalled, has the trailing edge of the blade directly facing into the wind causing the rotor assembly to not move at all. You'll want to fully stall a wind turbine in extremely high winds or for maintenance shut downs.

> **NOTE**
>
> Stall control and pitch control differ in that stall control diminishes the rotor blade's angle of attack reducing the blade's lift-to-drag ratio while pitch control increases the lift-to-drag ratio.

Active Yaw Control

Yaw control on a wind tower is the circular motion around the vertical axis of the tower parallel to the ground. The yaw control mechanism and gears are located in

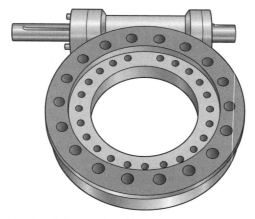

FIGURE 2-4 Example of a wind tower slew drive.

the nacelle frame at the top of the tower structure. This allows the yaw control system to move the whole towerhead to include the rotor assembly, turbine, and nacelle. The mechanics of the yaw system are a large ring gear (slewing drive) and bearing assembly **FIGURE 2-4**. A motor is used to drive the rotation movement of this gear assembly.

Wind energy is free moving and does not always come from the same direction. Active yaw control will rotate the towerhead, allowing it to move and chase after the changing wind direction. When a wind turbine syncs up with the wind, it maximizes the efficiency of the tower.

In excessively high wind conditions, the towerhead should be turned away from the forces of the wind. This turning of the towerhead a number of degrees out of the wind is called *furling* and results in a safer rpm of the rotor assembly. Furling allows the turbine systems to come back into alignment with their intended parameter of operation and reduces damage to its wind turbine.

However, furling can put the blades and the entire turbine assembly into different stresses, and it can cause unusual vibrations, depending on the design. This is one of many design considerations showing the connection between blade design and control systems design. Under more severe conditions the towerhead can be furled even further by turning a full 90 degrees to the wind and waiting for the rotation speed (rpm) to drop sufficiently enough to apply the mechanical brake controls.

NOTE

"To furl" means to wrap or roll something (e.g., a sail or a flag) in preparation for storage. The wind generation industry takes leeway with this definition by suggesting the yawing action of a towerhead is wrapping or rolling back on itself in preparation for shedding the wind energy. In a stretch of the imagination, the tower has gone into storage (furled sail or flag) as in a state of rest or in waiting.

Blade Pitch Control

An essential part of rotor assembly design is using blade pitch control for limiting power from the turbine to the power grid. Typically, wind speeds over 45 mph generate turbine electrical output too high for the central collection system of the wind farm (substations, transfer stations, and power grid) at which point the turbine must power back its supply.

This style of limiting power control (blade pitch control) requires both an atmospheric monitoring system and the rotor blades' mounting angle at the hub to have adjustable blade capability and pitch input. Variable blade pitch control requires a monitoring system for variable inputs of both wind direction and wind speed. This information comes from the **anemometer**, a mechanical wind speed and direction device, sitting outside on the nacelle. It also comes from a meteorological tower located at the most prominent **upwind** direction. This is the direction from which the wind is blowing in relation to the location of the wind farm facility.

NOTE

The meteorological tower ties into all the wind farm towers for overall control input to the wind farm development.

Normal operations of a variable blade pitch wind turbine will typically adjust the rotor blades a fraction of a degree while the rotor assembly is turning. This is done in order to keep the rotor blades at an optimum angle to produce the maximum output for all wind speeds until such time the wind speeds become too strong or erratic. Next, the electronic controller tells the blades to alter their pitch to become unaligned with the wind, slowing down the rotor blade rotation and holding the rotor assembly within the design parameters of the set speed for rotation.

Rotor Hub

The rotor hub is an essential part of the wind turbine. All of the combined forces experienced on the rotor blades (such as the rotational driving loads, gravity loading by blade weight, and the blade's weight) are felt at the rotor hub. Each of these loads acts upon the hub and is transferred to the mechanical drive train. While the hub is being acted upon by these loads, it is also receiving control input for adjustments to the rotor blade angle of attack.

The rotor hub is the most heavily loaded component on the towerhead. Because of this, it must receive operational power from external sources (outside the rotor assembly). If an emergency situation arises and the turbine monitoring system requires the blade to be pitched to its maximum, then the blade pitch mechanism must come into play over and above the blade pitch control system.

Blade Pitch Mechanism

Blade pitch mechanisms are typically hydraulically actuated to withstand the strong rotation forces at play. This is the most heavily loaded element of a wind

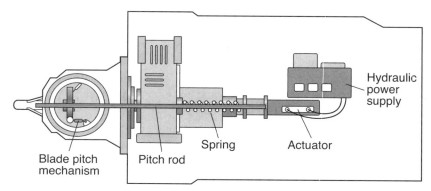

FIGURE 2-5 Hydraulic pitch mechanism with a central hydraulic spring-loaded actuator.
Adapted from Harrison, R., Hau, E., & Snel, H. (2000). *Large Wind Turbines Design and Economics*. West Sussex, England: John Wiley & Sons Ltd.

turbine with special heavy-duty pitch bearings and high-performance ring seals used at the blade root. Hydraulic actuation, which is noted for its slow, steady movement, is employed, as the rotor blades need to turn smoothly in response to varying wind conditions **FIGURE 2-5**.

The hydraulic blade pitch mechanism is spring loaded for operating in a fail-safe mode to power down the rotor assembly upon power failure. When loss of electrical control power occurs, this fail-safe mode will cause the hydraulic actuated spring to de-energize and revert to its natural state, which is collapsed. This actuation pulls the pitch mechanism rod into a position whereby the rotor blade is rotated to a stall position.

NOTE

Some manufacturers are experimenting with electric servomotors in which each blade can pitch individually. This allows for a more responsive control and possible cyclic pitching that could address wind shear on the blade loads.

Rotating Seals

When the rotor blade pitch angle is fixed and the blade is bolted in place on the hub there is no need for special seals between each rotor blade and the hub. A standard ring seal between the hub frame flanged joint and the blade flange is sufficient. However, on a variable pitch blade the seal requirements escalate due to the blade rotating on its longitudinal axis while nested into its hub support. This connection requires a special ring seal with a diameter ranging from five to seven feet.

A wind turbine rotor shaft seal is made of Teflon (PTFE) for its high performance, extremely low friction, high speed, and long service life. Because of its large diameter, the seal will have a stainless steel garter spring to provide uniform contact and a pilot leg for alignment. A heavy-duty anti-rotation leg of the seal will seat against the rotor blade, which is able to withstand shaft misalignment **FIGURE 2-6**.

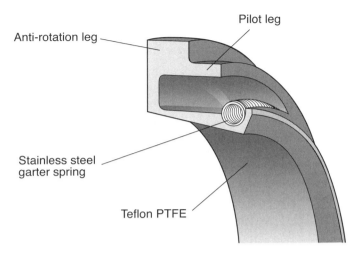

FIGURE 2-6 Cutaway section of large-diameter rotary seal.
Adapted from System Seals Inc.

Rotor Braking System

Safety regulations vary from country to country, but most countries require a rotor braking system. Regulations further require that redundancy be used in their safety procedures so both a primary and a secondary braking system is installed in all wind tower machines. The primary method of braking uses the aerodynamics of the rotor assembly already incorporated into the towerhead power control system. A secondary method is an actual mechanical braking of the tower mechanical drive train and the yaw shaft.

Purpose

In the event of a malfunction of a critical towerhead component or any circumstance of a potentially dangerous nature, the wind tower must stop the rotation of the rotor assembly immediately. It is a rotor braking system's function to bring the rotor assembly to an automatic full stop and thereby also stop the other towerhead systems fully. Also it is essential to have an over-speed protection procedure. To do this the turbine safety control system incorporates both an aerodynamic braking system (pitch and yaw control) and a mechanical braking system. For both of these systems to function reliably, they depend heavily on towerhead monitoring, the meteorological site tower, and turbine process controls.

Aerodynamic Braking

Aerodynamic braking systems have proven to be the best way to brake and stop the rotor assembly because they are extremely effective and safe. This system is already an established procedure with the rotor assembly via the stall and blade pitch power control. Typically when applied within a couple of blade rotations, the rotor assembly will have stopped itself.

The rotor braking system already monitors the wind conditions but as a safety measure it also ties into the towerhead monitoring and process controls for generator overheat temperature, excessive vibration levels, rotor over-speed, loss of power, loss of grid connection, and many other set points and monitoring inputs. This operating procedure is accomplished without any undue stress, wear and tear on the towerhead over and above what is the norm. Essentially the aerodynamic braking system consists of turning the rotor blades about 90 degrees along their longitudinal axis for an override of the current wind condition and pitch (angle of attack) of the rotor blade.

NOTE

Aerodynamic braking is applicable for pitch-controlled and active-stall controlled rotor assemblies only. In passive-stall controlled turbines where the rotor blades are stationary (or fixed), the rotor blade tips turn 90 degrees as a means for a safety braking system.

The blade pitch mechanism is engineered to fail in a fail-safe mode such as in the case of electrical power failure. Both the stall and pitch systems are designed to operate via this fail-safe mode. Activation of the rotor blade requires power. The lack of power, either hydraulic or low-voltage control power, will power down the brake mechanism and cause it to go into a mode of operation that is a safe state free from danger or risk. This is accomplished by spring loading the mechanism and having the spring revert to its natural state of compression when there is a loss of hydraulic or electric control power. Once the spring is in its natural compressed state, the pitch rod is realigned causing the rotor blade to turn 90 degrees. This faces the trailing edge of the blade directly into the wind in a stall position.

The most efficient and most safety minded way of stopping a large commercial wind turbine (for any reason) is the use of the aerodynamic braking system.

Mechanical Braking

The mechanical braking system gives a secondary (redundant) backup to the primary aerodynamic braking system as a means to fulfill safety regulations. It is rarely used as the rotor assembly, once pitched to 90 degrees, can move very little if at all against the weight and **torque** pressures already inherent in the towerhead design. Torque is the amount of pressure it takes to move the rotor. However, once a turbine is stopped, the mechanical brake functions very well as a parking brake to hold the system in a confirmed stalled state.

The mechanical brake is most typically a disk brake mounted on either the low-speed main drive shaft before the gearbox or the high-speed shaft between the gearbox and the generator. If it's mounted on the high speed shaft, it requires much less torque exerted for braking and therefore can be smaller and less expensive. However, if the brake is mounted after the gearbox it offers no protection to the gearbox as the low-speed main drive shaft still wants to rotate causing an out-of-sync meshing of main drive-shaft gears to the low-torque, high-speed gears. If this happens gears can be stripped and the whole gearbox destroyed.

The mechanical brake is to be used only after the aerodynamic brake system has been applied and brings the rotor assembly and subsequently the tower mechanical systems down to a reduced speed. The mechanical brakes would wear quickly if used to stop the turbine from full speed.

> **NOTE**
>
> Sometimes there may be an additional mechanical brake on the yaw system of the tower. This brake prevents any unwanted oscillations of the towerhead after the rotation for aerodynamic braking is complete.

Mechanical Drive Train and Nacelle

The towerhead of a horizontal wind tower is designed around two main support structures—the mechanical drive train and the nacelle.

The nacelle serves three purposes. First it provides a protective cover to the towerhead against atmospheric elements. Next it forms the overall frame and support structure for the towerhead components. Last and most important, it establishes the beginning of a structural **load path** from the towerhead to the tower itself. This is the path taken by load forces from one component to another and is continuous until such time as the path route terminates in the tower base at ground level.

> **NOTE**
>
> Load path is a fundamental principle of structural design. It designates the transfer of normal load forces and deviant load forces (friction, vibration, oscillation) to a more secure and less destructive location. This design principle uses the mechanical tie-in of loads to a continuous and common path, tying one load force into another and another throughout a structure. The path continues and does not end until such sequence is terminated at the base of the structure.

Two Functions of the Mechanical Drive Train

The mechanical drive train resides inside the nacelle, and runs the length of the enclosure from the rotor hub to the turbine generator. The components typical to a drive train are the large drive shaft, gearbox, mechanical brakes, and small drive shaft **FIGURE 2–7**.

The mechanical drive train has two functions. First it transfers mechanical power from the rotor assembly to the turbine generator at a rotational rate (rpm) compatible to the generator. However, the drive train's rotational forces, friction, bending moments, and vibration cause stress and mechanical fatigue if not dissipated. (A **moment**, in this context, is rotational force around an axis.) Hence, the second function of the drive train is its support structure. This support structure is designed as the beginning of a load path that transfers to the nacelle and then to the wind tower.

Mechanical Power to Electrical Power

The mechanical drive train functions to transfer the collected mechanical power of the rotor assembly to the turbine's electricity-generating generator. More specifically it will convert the large torque with low rpm (around 30 rpm) from the rotor

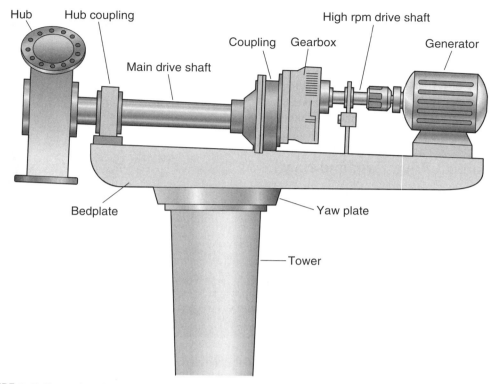

FIGURE 2–7 Illustration of mechanical drive train and components in relation to the nacelle support frame.
Adapted from Plantier, K., & Smith, K. M. (2009). *Electromechanical Principles of Wind Turbines For Wind Energy Technicians, 1st ed.* Waco, TX: TSTC Publishing.

hub large shaft to a lower torque yet higher rpm on the smaller shaft. This smaller shaft is more in line with the generator rpm (usually running at 1,000 to 1,500 rpm). There are numerous drive-train configurations that can accomplish this; each has its own advantages.

The drive-train gearbox is the key to how this changeover from low to high rpm and high torque to low torque takes place. The initial high torque main drive shaft originating from the rotor assembly connects to the gearbox via the *planet carrier.* This planet carrier is dynamically balanced with three small shafts (in line to the main shaft) each having its own smaller *planet gear.* Rotating in the center of these three planet gears is the *sun gear.* This whole assembly is enclosed in a *ring gear.* It is these gears and their respective configuration that changes not only the speed ratio of the main shaft to the small shaft but the torque required to rotate the smaller shaft **FIGURE 2-8** .

The diameter of the gears inside the gearbox is the determinate factor for achieving the amount of change in speed and torque ratios. Gear ratio is the relationship of

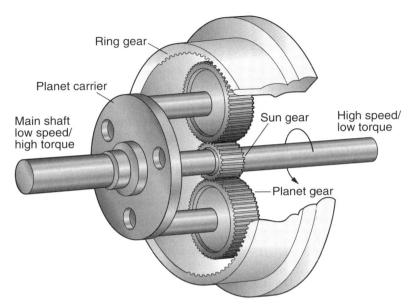

FIGURE 2–8 Typical planetary gearbox used most often in the wind energy industry.

Adapted from Plantier, K., & Smith, K. M. (2009). *Electromechanical Principles of Wind Turbines For Wind Energy Technicians, 1st ed.* Waco, TX: TSTC Publishing.

the teeth on one gear to the teeth on another whereby the two mesh together—one turning the other at a precise rate of revolution. Gear teeth and gear diameter determine the ratio of change and hence the resulting rate of speed (rpm) and the torque (ft.-lb.) to match that of the turbine generator.

Load Path From Turbine to Tower

The mechanical drive train not only inherits several load forces from the rotor assembly (torque, thrust, weight, aerodynamic) but also its own inherent load forces of rotation (vibration, oscillation, friction).

Starting with the rotor hub shaft, the mechanical drive train has on its operational load path the high- and low-speed shafts, gearbox, braking mechanism, couplings, bearings, and all other assortment of low voltage control panels, process controls, and monitoring devices. These components fail often due to torque-related stress. Manufacturers have developed huge, costly ring gears and bearings to dispel just such loads onto a purposeful load path.

The gearbox typically sits within a trunnion-style mounting that is specifically designed to transfer the residual vibration and oscillation to the drive-train chassis. Additional spring and rubber mounts also dispel torque and load forces from the shafts to the shaft coupling onto the chassis frame.

NOTE

Gearbox design varies greatly from manufacturer to manufacturer, but the planetary gearbox design is the most common on wind turbines. Parallel and epicyclic are two other gearbox designs in use by turbine manufacturers; each has its own advantages.

Directly beneath the mechanical drive-train frame chassis is the nacelle bed-plate. This plate is the transfer structure of the drive train's deviant load stresses, such as friction, vibration, oscillation, to the nacelle support structure. The nacelle accepts these forces in its frame structure and passes them along to the tower itself. The tower column is designed to offload these forces at its base into the tower piling and then the foundation platform. Ultimately they are transferred to the surrounding ground of gravel-bedded compacted earth.

Drive-Train Configurations

Drive-train configurations are many and varied from manufacturer to manufacturer, each having their own features and benefits. However, they all follow a general configuration of transferring high torque with low rpm from the rotor assembly to the low torque high rpm required by the turbine generator.

As discussed previously in the chapter, the shorter a load path not only means less weight but less bending moments. Similarly, the shorter the drive-train shaft, the less probability of fatigue and stress.

Classical Modular Drive Train

In a classical modular drive-train layout **FIGURE 2-9**, all the individual components mount separately and directly to their respective shaft (hub, main bearing, main shaft, gearbox, and generator shaft) and to the frame chassis. This makes for a longer drive train, adding more weight, more rotational bending, and more load forces on all four drive shafts. The three couplings between each shaft are designed to offload load path forces via bolt design, compression seals, and mountings for additional load path transfer capacity to the chassis frame **FIGURE 2-10**.

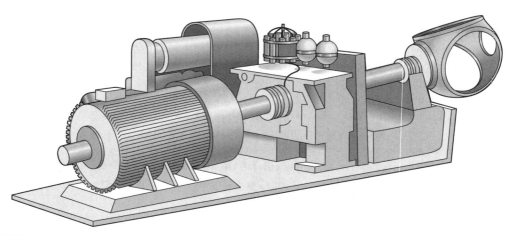

FIGURE 2-9 Example of classical modular drive-train configuration.

Adapted from Plantier, K., & Smith, K. M. (2009). *Electromechanical Principles of Wind Turbines For Wind Energy Technicians*, 1st ed. Waco, TX: TSTC Publishing.

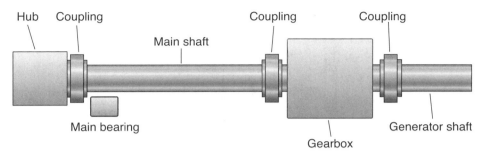

FIGURE 2-10 Example of classical modular drive-train configuration, showing couplings between each shaft.
Adapted from Plantier, K., & Smith, K. M. (2009). *Electromechanical Principles of Wind Turbines For Wind Energy Technicians, 1st ed.* Waco, TX: TSTC Publishing.

The gearbox uses a trunnion mount for suspension between the main drive shaft and the generator shaft and above the chassis frame. The trunnion mount itself acts as a load path stabilizing and transferring rotational vibrations from the gearbox to the chassis frame. Specialized heavy-duty bearings also act to lessen the rotation load forces and relay them into the load path of the chassis frame.

> **NOTE**
>
> The drive-train mechanical brake can be located either before or after the gearbox. If mounted on the main shaft, it helps protect the gearbox. However, mounting on the smaller shaft requires less size and weight and involves lower costs. Both styles are used.

Compact Drive-Train Concepts

Short or compact drive-train configurations are showing up more and more in wind tower designs. With the advent of larger wind turbine designs comes the desire to lessen the overall weight load. This can be accomplished by shortening the length of the main drive shaft, thereby allowing for a smaller nacelle.

A three-point suspension **FIGURE 2-11A** of the drive train allows for a free-standing front bearing while the rear shaft bearing is integrated into the gearbox. This layout requires only two mountings in total as compared with the mounting of each individual component of the classical drive train.

To further combine the load carrying dimension, the configuration shown in **FIGURE 2-11B** carries the entire weight of the rotor assembly and the main shaft along with the bearings of both shafts in the gearbox. The generator is then mounted directly to the gearbox for a much compressed drive-train length. The gearbox then acts not only as the housing for its gear wheels, but as the frame support of all the other drive-train components as well. This is a very straightforward and compact design. Only a parallel gearbox can be used in this configuration, however. Parallel gearboxes are not as popular with turbine designers because they are not as efficient as the planetary gearbox and cannot operate at the extremely low speeds of a planetary gearbox.

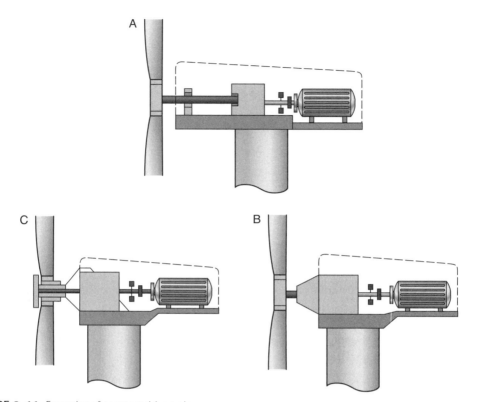

FIGURE 2-11 Examples of compact drive trains.
Adapted from Harrison, R., Hau, E., & Snel, H. (2000). *Large Wind Turbines Design and Economics*. West Sussex, England: John Wiley & Sons Ltd

A third sample configuration **FIGURE 2-11C** uses the nacelle frame itself as a main drive train and chassis frame mounts. The load path from the rotor assembly with its thrust, weight, and aerodynamic loads is directly channeled through the rotor hub bearings to the static nacelle frame. This makes the overall drive-train load path short and direct; however this approach has proven to be more expensive.

Drive-Train Dynamics

Drive-train dynamics and reliability go hand in hand. The mechanical drive train not only inherits all the load forces and resonant frequencies from the rotor assembly (mainly the rotor blades), it also has its own inherent resonant frequencies; the gearbox cogwheels being a primary source. If these resonant frequencies of vibration and oscillation are left unchecked, they will eventually facilitate an increase in maintenance and repair to the tower.

Variable speed operation and stall-regulated wind turbines promote more vibration and oscillation tendencies than a fixed-blade rotor assembly. **Damping** of the wind turbine through rotor speed control is the first line of defense. The ability to reduce drive-train torque fluctuations is very beneficial but not enough. You still have the collective lead-lag rotational stress to the shaft of the rotor blades to contend with. Add to this the drive-train distortions from both blade wind shear and the lateral tower loads. All these design components compound and contribute to the inherent **resonant frequency** of the drive train. Resonant frequency is the natural inherent low-level vibration produced when a system is in its operational state.

The length of the drive train is also a consideration. The longer the drive train is, the lower the resonant frequency. This length then becomes an issue as drive-train technology is striving to become more compact. Damping of the drive train mechanically through its individual components becomes a must. Rubber and spring-actuated mountings, couplings, impact seals, and trunnion-mounted gearbox all contribute to dispersal of vibration and oscillation to the towerhead load path.

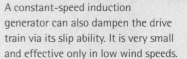

NOTE

A constant-speed induction generator can also dampen the drive train via its slip ability. It is very small and effective only in low wind speeds.

R&D designers have been using vibration monitoring and noise radiation calculations by computer-aided instrumentation to fine-tune these drive-train dynamics. These process control systems allow for pro-active damping of the drive train and off-site monitoring for both operational sequencing and preventive maintenance and repair.

Direct-Drive Drive Train

A direct-drive drive train can be used if the rotor assembly is fixed (no variable pitch/stall or power control) with the rotor blades designed to stay within a fixed range of speed compatible with the generator. A generator used with a direct-drive drive train has a unique design intended specifically for this style of drive-train configuration **FIGURE 2-12** . Also, in this configuration, there is only one main drive shaft.

The rotor assembly and drive train are designed to hold steady at an rpm of 60 hertz. If the generator is operating at this rate, which is normal, then the drive train has no need for a gearbox.

The mechanical drive-train gearbox cost is substantial in comparison to the overall drive-train costs, and these gearboxes are notorious for failures that lead to frequent system downtime and expensive repairs. Wind tower design is driving gearbox technology to seek out new methods to lighten the nacelle load, not just from the gearbox but also along with its influence on all the mechanical components around it, before and after they interact with the gearbox.

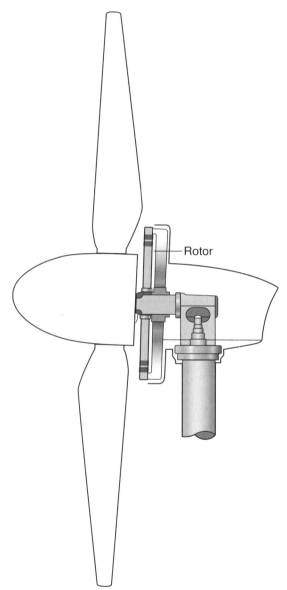

FIGURE 2-12 Direct-drive generator, whereby lengthy drive shaft and gearbox are eliminated.

Adapted from Harrison, R., Hau, E., & Snel, H. (2000). *Large Wind Turbines Design and Economics*. West Sussex, England: John Wiley & Sons Ltd.

EXPERIMENTAL ROTOR DESIGNS

As wind-generated power grows in popularity, so too does the need for flexibility in the location of towers and their design.

Traditional horizontal wind turbines rotate their towerhead to follow the change in wind direction. New designs allow for wind collection from all directions. The "QR5" and "Windspire" are both such vertical turbines. They not only are smaller, they offer an overall futuristic style of design with their unique blade designs **FIGURE 2-13**.

You can also mount wind turbines on the top edge of a building. The Architectural Wind turbine, developed by AeroVironment, is a small turbine that takes advantage of the wind lift created at the side of a building. As the wind hits the building and resistance creates an updraft, the wind accelerates up the building wall. The Architectural Wind turbine hanging out over the building's top edge catches these accelerated winds before they disperse over the top of the building.

WhalePower Corporation created a blade design based on a humpback whale's fins. These tubercle blades have been shown to increase electrical production by 20 percent.

FIGURE 2-13 What the future holds for wind turbine designs.
© iStockphoto.com/RFStock

CHAPTER SUMMARY

The mechanical system of a wind turbine may look to be no more than a support structure, but its dynamics are the true backbone of wind turbine power generation. Materials science, computer science, aerodynamics testing, and component innovation all take their places at the mechanical design drafting table.

KEY CONCEPTS AND TERMS

Anemometer

Coefficient of power (C_p)

Damping

Horizontal axis

Load path

Moment

Resonant frequency

Tip-speed ratio

Torque

Upwind

CHAPTER ASSESSMENT: WIND TURBINE TECHNOLOGY AND DESIGN CONCEPTS, PART 1

1. A rotor assembly can have only three blades.
   - ❏ A. True
   - ❏ B. False

2. What is the Betz theoretical maximum power coefficient (C_p) value of a horizontal wind turbine?
   - ❏ A. 0.45
   - ❏ B. 0.35
   - ❏ C. 0.59
   - ❏ D. 0.50

3. On a horizontal wind turbine, the ratio of _____ to the amount of _____ is the power coefficient.

4. The tip-speed ratio of a rotor assembly is the ratio between the rotational speed of _____ and the actual _____ of the wind.

5. Resonant frequency for the mechanical drive train can be:
   - ❏ A. friction.
   - ❏ B. vibrations.
   - ❏ C. oscillations.
   - ❏ D. All of the above

6. Blade flexibility allows for improved energy capture.
   - ❏ A. True
   - ❏ B. False

7. Designing a rotor blade with flexibility allows for which of the following?
   - ❏ **A.** Angle of attack for the blade made larger
   - ❏ **B.** Shedding wind gusts
   - ❏ **C.** A better tip-speed ratio
   - ❏ **D.** Tie in at the hub made easier

8. Stall control diminishes the effects of the wind slowing the rotor assembly down, while pitch control fluctuates the rotor blade to maximize efficiency.
   - ❏ **A.** True
   - ❏ **B.** False

9. Each rotor blade has the capability of rotating around what axis?
   - ❏ **A.** Vertical
   - ❏ **B.** Horizontal
   - ❏ **C.** Longitudinal
   - ❏ **D.** All of the above

10. The primary rotor assembly braking system is _____ braking, and the backup braking (redundant) system is _____ braking.

11. What active power control system rotates around the vertical axis of the wind tower?
   - ❏ **A.** Stall control
   - ❏ **B.** Yaw control
   - ❏ **C.** Blade control
   - ❏ **D.** All of the above

12. What produces the most stress and fatigue on the mechanical drive shaft?
   - ❏ **A.** Generator
   - ❏ **B.** Main shaft
   - ❏ **C.** Gearbox
   - ❏ **D.** Trunnion mount

Wind Turbine Technology and Design Concepts, Part 2

WIND TURBINES BECOME a commercial venture when you connect them to an existing power grid and sell the wind-generated electricity for profit. This is all possible through the electrical concepts of wind turbine technology and wind tower design.

While each wind turbine, generator, and tower manufacturer may promote its products to developers as proprietary designs and methods of production, they all must also adhere to two very distinct fundamental concepts: mechanical and electrical. This chapter covers all electrical concepts up to but not including the technical field of process controls. Process controls encompass all of the instrumentation and respective wiring schematics that both the mechanical and the electrical systems require in order to be monitored and interfaced. This is a complex subject—a full book on its own—and will not be addressed here specifically but made note of and referenced as a peripheral subject.

Chapter Topics

This chapter covers the following topics and concepts:

- The three most prominently used generator styles: inductive, synchronous, and direct-drive
- The transfer of electrical energy from the wind tower to a utility power grid

Chapter Goals

When you complete this chapter, you will be able to:

- Describe the features and benefits of inductive, synchronous, and direct-drive generator systems
- Relate how a generator converts rotational mechanical power to electricity
- Explain how low-voltage turbine electricity is stepped up, increased in capacity, and transferred to a power grid

Electrical System

The electrical component aspect of a wind tower starts with the generator located in the nacelle at the towerhead. The generator initiates the beginning of the wind tower electrical system and comprises the heart of wind power design **FIGURE 3-1**. It is where the wind turbine's mechanical power converts to the most versatile power source ever invented—electricity.

With electricity, you have the means to transport power over long distances. The most extreme energy requirements of industrial production or household conveniences can be met via the conveyance of electrical power through flexible and malleable wires that form a nation-wide web of power lines that form an electric transmission grid of massive proportions.

To understand the heart of this massive system you will need to understand the general workings of the towerhead generator. The generator mounts in line with the towerhead's main drive shaft and is coupled to the low-torque, high-speed, small drive shaft via its own steel alloy shaft.

There are numerous configurations to this main drive shaft coupling sequence. However, all towerhead configurations adhere to the same concept: the generator should be coupled last to the rotational mechanical power of the wind turbine.

Electricity is created specifically at the juncture of the generator rotor and stator **FIGURE 3-2**. The stator consists of a steel housing (frame) allowing for a hollow cylindrical core inside, in which the rotor resides and rotates clockwise with the rotational motion of the generator shaft. There are numerous generator design approaches, all adhering to the basic principles of electromagnetism as applied to electric generators. Common to all generator design approaches is the excitation via magnetic charging of a *field coil* (often called field winding) whereby an electrical charge is produced.

Generators used in a wind towerhead require a slightly different operating sequence from typical generators (e.g., hydroelectric, bio-fuel, oil/coal) attached

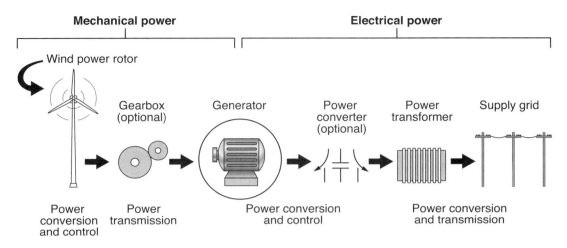

FIGURE 3-1 Main components of a wind farm, with the generator at the heart.

Adapted from Blaabjerg, F., & Chen, Z. (2006). *Power Electronics for Modern Wind Turbines*. 40 Oak View Drive, San Rafael, CA: Morgan & Claypool publishers.

to commercial power grids. This is due to the fluctuating energy source of the wind, which causes the fluctuating mechanical power (i.e., torque) of the rotating drive shaft.

Generator capacity is determined by the towerhead rotor assembly. The generator capacity must correspond with the size of the rotor assembly swept area. A generator adapts a marginal amount to the variances in load, but not so much as to overload its capacity. Generators of both high speed and low speed comprise the two approaches used for converting rotating mechanical energy to electrical current. High speed (1,000 to 1,500 rpm) generators are either induction or synchronous; low speed (20 to 40 rpm) generators are direct drive. There are also variable speed systems having advantages and disadvantages using both induction and synchronous-style generators.

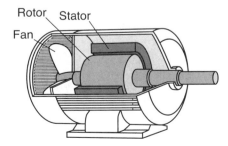

FIGURE 3-2 Cutaway drawing exposing rotor and stator assembly inside a generator.

Adapted from Manwell, J. F., McGowan, J. G., and Rogers, A. L. (2009). *Wind Energy Explained: Theory, Design, and Application*. West Sussex, England: John Wiley & Sons Ltd.

Induction Generators

An induction generator is basically an induction motor whereby the **slip** is negative. Slip rings on the generator shaft are the interfacing device that allows bushings to transfer electrical flow between the rotor and the stator FIGURE 3-3 . Negative slip takes place when the generator rotor speed is slightly ahead of the magnetic field in the stator winding. This is called **flux** and allows electrical flow to be created rather than like a motor where electricity is supplied for use to drive a mechanical device.

Induction generators are the rule at present in commercial wind turbine design, as they offer the best cost rewards with rugged construction in comparison to synchronous generators. They are most typically constructed in a closed form and require a heat exchanger for cooling. The advantage of induction generators is they allow variable power control turbine systems to operate at a more constant speed.

There are currently three approaches most commonly used for induction generators:

- Cage rotor—Also referred to as a squirrel cage. This is where the generator directly connects to the AC power line and operates at a fixed speed (with a 1 to 2 percent variance). The frequency of the power grid determines rotational speed of the generator.
- Wound rotor with slip control—Allows for variable power control (e.g. pitch, stall, and yaw) dynamics as the generator's use of slip control varies its rotor's resistance, widening the operating speed ability within a fixed range.

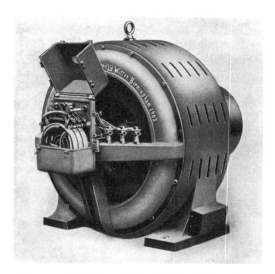

FIGURE 3-3 View of a generator's internals, from left to right: casing cover (open), rotor, and slip ring. The area behind includes the generator housing and shows the stationary stator with a hollow cylindrical core.

© VINTAGE POWER AND TRANSPORT/MARK SYKES/Alamy Images

- Doubly fed—Introduction of a power converter provides **vector** (magnitude and **phase** angle) control of the rotor circuit current. This widens the operating speed ability to match that of the turbine dynamic power control conditions. It is the power converter's job to then sync up with the power grid frequency.

Synchronous Generators

The **synchronous** generator operates at a specific fixed speed matching that of the grid frequency. If the mechanical torque being produced by the turbine rotor assembly becomes too great, the magnetic forces of the synchronous generator will resist and stay at its designated speed.

A sudden large gust of wind can put major stress on the generator due to this inherent design feature. A power control system must be used for the wind turbine rotor assembly to synchronize its generated frequency with the grid frequency. Because of this, synchronous generators are limited almost entirely to variable speed systems.

Variable Speed Systems

Both induction and synchronous generators can be used on variable speed systems **FIGURE 3-4**. A variable speed system incorporates external power

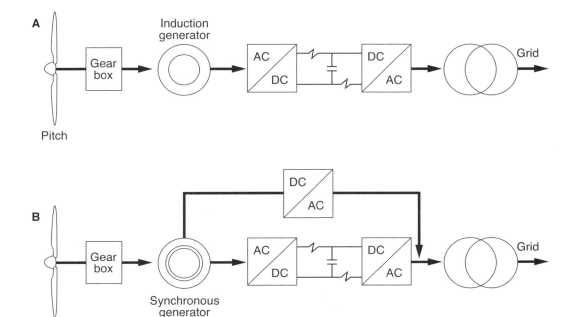

FIGURE 3-4 Full-scale power electric converter system showing inductive (A) and synchronous (B) generators.
Adapted from Blaabjerg, F., & Chen, Z. (2006). *Power Electronics for Modern Wind Turbines*. 40 Oak View Drive, San Rafael, CA: Morgan & Claypool publishers.

electronics used to connect the generator frequency to grid frequency. This system allows the generator to operate within its highest efficiency without having to put stress on its own components to handle the fluctuation in torque from the wind turbine.

A variable speed system also has the advantage of increased energy yield due to the generator being able to operate at its highest efficiency throughout a wide range of wind speeds and load fluctuations. The high cost of needing to convert the variable frequency via a power convertor is a disadvantage and requires an AC-DC-AC conversion link of electronic controls.

Low-Speed Direct-Drive Generators

As the wind generation industry looks to increase the magnitude of its output, so too must the designers look to individual components for increased output. Consequently, low-speed direct-drive generators are being redesigned for greater output capacity at lower generator rpm speeds. Low-speed direct-drive generators operating at 20 to 40 rpm are already in production. Researchers are studying other low-speed direct-drive generators in some detail with results showing 4 megawatt (MW) output with possible up-scaling to 10 MW.

Two factors determine the size of a low-speed direct-drive generator: rotor diameter and electromagnetic pole size. The generator rotor diameter increases due to shaft torque increase from the slow rotational speed (20 to 40 rpm) allowing for better rotor leverage. Also, there is an increase in electromagnetic pole size to give an increase in frequency of the generator. Companies exploring this approach want to minimize generator size while maintaining a frequency, typically 60 Hz, that matches the power grid.

Transfer of Wind Power to the National Grid

Commercial wind farms differ from conventional generating plants such as hydroelectric dams, fossil fuel (e.g., oil and coal fired), and nuclear generating plants. While their objectives remain the same, the manner in which they deliver electricity does not. Conventional generation facilities typically comprise a handful of large-scale generating units all housed in close proximity to one another in one powerhouse building.

In contrast, a wind farm can encompass anywhere from a handful of towers up to hundreds of towers located across a vast scope of land—some not even in sight of each other and each housing its own generator. A wind turbine generator is also quite small relative to conventional powerhouse generators where all their cumulative power is housed in one structure **FIGURE 3-5**.

The capability of a wind farm to collect each individual tower's generated power from numerous and far-reaching locations to a central, cumulative load makes for a power configuration of a different sort, unique unto itself **FIGURE 3-6**.

A   **B**

FIGURE 3-5 Modern power plant steam generator (A) compared with a relatively small wind generator (B), housed separately in each of the many wind towers on a wind farm.

Courtesy of the NRC

© TFoxFoto/ShutterStock, Inc.

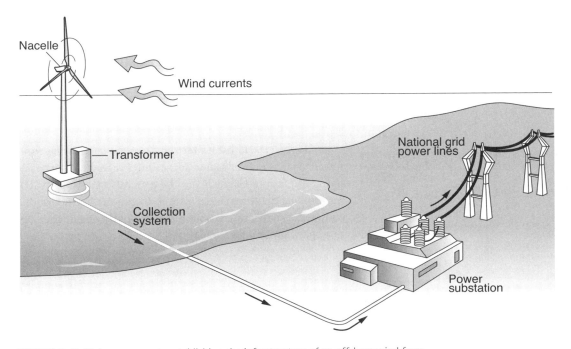

FIGURE 3-6 Main components establishing the infrastructure of an offshore wind farm.

This infrastructure configuration encompasses several uniquely different electrical networks:

- The wind tower itself acting as an electrical raceway
- A collection system of networked power lines
- A central collection and monitoring substation
- Interfaces into an already established power grid for commercial distribution

Tower

Delivery of electricity (e.g. low-voltage 400 to 690 V) from the towerhead generator to ground level makes use of the tower itself, which can be considered one large electrical raceway. Most commercial towers manufactured today are steel-plated structures, segmented and coupled together one segment on top of the other in diminishing diameter as the tower rises. While the hollow interior of the tower houses the access ladder to the towerhead and acts as structural load path for the towerhead to the ground, it is also perfect for electrical cable routing.

The exact means of routing the towerhead cable down through the wind tower depends on each tower manufacturer, but most often it is via a cable tray system. The cable tray system runs the length of the tower providing a secure pathway for all towerhead electrical cable. It forms a structural system where the cable is securely fasten, supported, and protected. The tray frame is considered a structural component of the tower facility electrical system.

Wind tower designers, always on the lookout for cost rewards and faster means of erecting a tower, have developed a pre-installed segmented cabling system. This system makes use of factory pre-mounted cable clamps with pre-assembled cable for each tower segment. Running tower cable at the factory eliminates the laborious operation of pulling cable through the tower onsite after it has been erected. This allows for reduced construction time without any structural tradeoff or technical disadvantage.

> **NOTE**
>
> Some wind towers allow room for housing the GSU inside the tower structure itself.

The towerhead cables terminate at the base of the wind tower into the top of a junction cabinet seated on the tower base foundation. Cables from the generator step-up (GSU), just outside the tower base, terminate into the bottom of this cabinet after running through a horizontal concert raceway just below ground level **FIGURE 3-7**.

Generator Step-Up (GSU)

The **generator step-up (GSU)**, located at the base of the wind tower, in actuality is a step-up transformer, and the first voltage step-up from the tower generator. (To **step up** voltage is to increase its magnitude.) The GSU will increase the tower low-voltage of 400 to 690 volts to a medium-voltage of somewhere between 1.5 and 35 kV for transmission to the wind farm substation via a collection system.

FIGURE 3-7 Wind tower base with generator step-up (GSU) off to the side.
© iStockphoto.com/pedrosala

The other more important reason for having a GSU right at the tower base is that low-voltage lines tend to have high resistance losses, and the longer the cable run the more losses will be incurred. At higher voltages with reduced current the reactive power losses due to line resistance are decreased. The voltage coming from the towerhead is only 400 to 690 volts and very susceptible to transmission losses.

Transmission and distribution line losses constitute the difference between the power produced and what is available for use. Losses in power transmissions can be significant. In the United States, these losses were calculated to show an estimated 6.5 percent in 2007. The ratio of real power (power available to a load) to reactive power is called the power factor. Reactive power represents line losses due to inductance and resistance, neither of which transmits power to a load.

Because of the GSU's size and weight, the closest location for this step-up transformer is at the base of each tower. This can be either inside the tower or just outside; it makes no difference, as long as the short cable run requirement is met. The GSU is also needed to increase the voltage for the longer transmission runs not just between towers, but to the central collection substation. These runs can be many miles and the lower voltages would create too great a resistance power loss.

Power meters and main circuit breaker are mounted inside the GSU cabinet for interface with the wind farm instrumentation and control (I/C) system. The GSU allows for the first protected space away from the tower for monitoring of the tower performance. Also, it provides a service area for isolation of the tower.

NOTE

Transmission of power from a wind tower to the power grid can take place at low, medium, or high voltage. Offshore wind farms use high and extra-high voltage. Most wind farm transmissions today are at medium voltage (1.5–35 kV).

Collection System

All of the wind-generated electrical power from each and every tower in a wind farm needs to be collected into one location before transmission to the power grid can take place. The extensive configuration of a collection system fulfills this requirement and allows for control and monitoring of each tower from a central operations and maintenance (O&M) building on site.

The GSU stepped-up power line leaves each individual tower as an underground feeder line with a medium voltage of 1.5 to 35 kV. This feeder line begins the collection system infrastructure of the wind farm. As one feeder line departs a tower it will couple up with another tower feeder line closest in proximity via a splice box. This is a freestanding field cabinet that resides usually at the junction of two or more tower access roads. As more towers are spliced into the collection system a field junction box will be used.

Depending on the layout of the wind farm, each wind tower will be brought online with this system until a group is formed. Most often this is done via groups of towers in proximity to each other; each group receiving its own distinctive recognition name usually as a nomenclature number. At this point, the collection system of individual groups will come together on a distribution line. The line voltage does not change; it is still a medium voltage of 1.5 to 35 kV, and in most cases still runs as buried cable. There are occasions when wind farms use overhead collection lines rather than buried cables, or some combination of the two.

Advantages of buried cable:

- It is less subject to severe weather conditions (e.g., lightning, wind, and freezing rain) than overhead lines. Buried cable poses fewer hazards to low-flying aircraft and wildlife.
- It benefits from the greater shielding effect provided by the surrounding earth, not only for the safety to the cable itself but also to minimize health concerns about the emitted electromagnetic field (EMF).
- It requires less installation strip area (1–10 m), whereas overhead lines require a space 20–200 meters wide and high enough to accommodate the height of utility poles. Overhead lines may also require removal of vegetation.

Disadvantages of buried cable:

- It is more expensive to install than overhead lines and much more costly to maintain if it requires digging up. Overhead lines are more accessible and easier to repair.
- The reactive power of a buried cable produces large charging currents, which makes it more difficult to control voltage.

At various intervals within this collection system, additional equipment such as capacitor banks and switching devices are strategically placed. These devices enhance and control power flow as a means to reduce resistance losses for stabilization of the collection system voltage.

Substation

The wind farm substation **FIGURE 3-8** or transfer station, as it is oftentimes called, constitutes the final point-of-relay between the wind farm infrastructure and an established power grid system. The main function of a substation is to step up the collection system's medium voltage (1.5–35 kV) to the high voltage (115–600 kV) required of the particular power grid for intended connection.

Where individual wind towers may be miles away and in varied locations, the substation resides as close to the intended power grid connection as possible. It is an outdoor facility with perimeter fencing encompassing many banks of transformers as well as switching, protection, and control equipment. The high-voltage switching allows collector lines to be connected and isolated from the power grid and for fault clearance or maintenance.

The metallic fence surrounding a substation is heavily grounded to protect against any high voltages that could occur during a fault in the facilities network. Earth faults (or current flowing to the ground) are extremely hazardous and pose a threat of electrical shock.

Substations vary in size and range from simple to complex. A small substation can be nothing more than a switching station of busways and circuit breakers. Large complex substations physically can entail a compound as large as a city block. In addition to the standard switching gear and transformer banks a substation of this size will also have monitoring and control systems. The wind farm

FIGURE 3-8 A fenced wind farm substation of medium size.
© Pedro Salaverria/ShutterStock, Inc.

O&M building usually is situated close by for field monitoring, control, and operation reporting of the wind farm facilities.

Metering the substation power output is of vital importance as this constitutes the wind farm income and profit-generating side of the wind power process.

National Grid Integration

Power from the substation transmits through overhead power lines when it leaves the wind farm to connect with a commercial power grid. High-voltage power transmissions use buried cable only in sensitive locations or urban areas. This causes a substantial cost increase. To reduce power losses in long-distance transmission, the substation power is transmitted at high voltages (110–765 kV).

The location in most need of power does not always reside where wind generation is most abundant. However, once a wind farm taps into a power grid it becomes available to be transported throughout a national power grid system FIGURE 3-9 . The current US commercial power grid has three major networks: the Western Interconnection, the Eastern Interconnection, and the Electric Reliability Council of Texas (ERCOT).

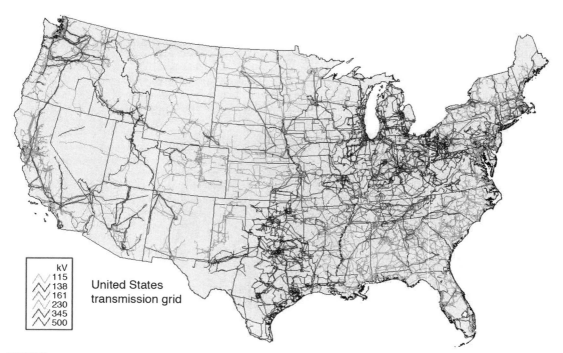

kV
115
138
161
230
345
500

United States
transmission grid

FIGURE 3-9 Power grid transmission line configuration throughout the United States, classified by voltage rating.
Courtesy of FEMA

In order for this large-scale integration to work efficiently, codes addressing minimum requirements have been established. As with any interface into an already established system, meeting these code criteria gives the assurance of continuity and stability.

The following are basic requirements to be addressed by a commercial wind farm in order for smooth integrations into the national power grid:

- Frequency and active power control—US national power grids operate on a standard frequency of 60 Hz. Due to the fluctuation of wind energy, the variable power control of the rotor assembly at the towerhead becomes a very critical system. It is the source of control; and, along with the generator power converter, it facilitates operation of the generator at a continuous frequency of 60 Hz.

- Reactive power control—Control of reactive power flow is needed for the reduction of line losses and stabilization of system voltage. The oscillation of energy stored in capacitor banks and consumed in inductive components is reactive power. It is power that does not contribute and gives no value toward real power (usable power) and therefore must be controlled.

- Short circuit power level and voltage variations—the ability of the wind farm collection system to absorb disturbances of short circuit and voltage variations is critical. A *strong collection system* and/or *weak collection system* refers to the strength of a wind farms' power-level capabilities. The wind turbine and collection system strive to maintain a steady-state voltage until it reaches the substation **point of common coupling (PCC)**.

- Voltage flicker—Most common and noticeable example of voltage flicker is when an electric light bulb momentarily flutters. Slight but perceptible variation of voltage caused by fluctuation of power generation may cause voltage quality problems. Rapid generator and capacitor switching may also result in line-voltage variation.

- Harmonics—Harmonics are a distortion of the sinusoidal waveform and can be present in voltage or current, or both. Most often harmonics are perceived as audible static on a line, easily removed by filters. Depending on the harmonic order of magnitude, it can also cause damage with increased currents and possible destructive overheating in capacitors. The power grid requirements should be in place and maintained before power is received by a national grid.

- Stability—A loss of power production at the wind farm reflects immediately on the power grid, and disconnection of the wind farm from the power grid only serves to escalate the situation as a brownout or power imbalance would then occur. Short circuits and faults are the primary culprits in large momentary power imbalance. Wind farms are required to have in place the ability to ride out such transient disturbances.

VERTICAL AXIS WIND TURBINES

Vertical axis wind turbines are not as common in the commercial wind industry as the horizontal axis wind towers, but they have their advantages. They never need to chase after the wind, because their rotor axis is vertical to the ground. The rotor blades are situated to capture the wind from any direction, thus eliminating the need for yaw control. The conventional design of a vertical axis turbine looks like an eggbeater whisk, each blade coming into contact with the wind from any direction **FIGURE 3-10**. This design also eliminates the need for a tower support structure.

Not needing a tower support structure gives a cost reward in itself. Lack of a tower also allows for maintenance and repairs of the turbine within easy reach of the ground, but it is also a drawback. The higher a wind turbine tower, the more likely a consistent and forceful wind velocity will be encountered for more efficient commercial electrical power output.

The aerodynamics of a vertical axis to a horizontal axis turbine are quite similar; they both use airfoil dynamics. But this is where the similarities stop. A horizontal rotor blade has the major portion of its blade designed for a variable angle of attack and therefore a good lift-to-drag ratio. Only about 25 percent of a vertical axis blade gives a strong lift and driving force along its blade even under optimum conditions. The horizontal turbine not only has a higher tip-speed ratio, but it extends over a greater ratio spread.

In an overall comparison, the vertical axis arrangement has more disadvantages than the horizontal axis at comparable power levels so they have not been used on large-scale commercial wind farms to date.

FIGURE 3-10 Vertical axis wind turbine featuring eggbeater-style rotor blades.
© Photos.com/Thinkstock

DO WIND FARMS AFFECT THE WEATHER?

Environmental engineers are starting to ask the question: "Do large wind farms affect the weather?" They are performing extensive studies and computer climate modeling to find the answer. Their studies make for some very compelling discoveries.

Large commercial wind turbines, when grouped together in a wind farm, are believed to disrupt the natural airflow of the region. The rotor blades of each wind tower create wind turbulence downwind of each towerhead setting up vortices similar to that of a wake made from a boat propelling across the water.

These downwind turbulences create a strong mixture of heat and moisture. Such a mixing effect near the ground causes the nearby land surface to become warmer and drier, which means the need for irrigation will increase to ensure the survival of the crops. During the daytime, natural turbulence (a mixing of the layers of the atmosphere) is already present. It is the predawn hours that hold the most concern. This is a time period of little or no natural turbulence. But because wind towers operate around the clock, it is believed a large wind turbine can raise temperatures by about four degrees Fahrenheit during these hours. At slow rotor speeds this turbulence effect does not exist. Since it is an issue only during the predawn hours, many wind farms use their power control systems to power back during these hours. There is a loss of revenue, but not so great as not to warrant the effort.

Another concern is the possibility that the turbines' rotating blades may redirect high-speed winds from above the ground to ground level, thus boosting evaporation of soil moisture. In larger-scale studies, it was found the average wind speed passing through a wind farm could be lowered by 5.5 to 6.7 miles per hour immediately downwind.

CHAPTER SUMMARY

Wind generated electricity is unique due to the fluctuations of the wind and the relatively small-sized generator housed in each individual wind tower. The wind generator is the heart of a wind farm development; it converts mechanical power to electricity and produces it in a consistent state so it's viable for conveyance to a national power grid.

Wind towers spread out over vast landscapes and pose another challenge in the collection of wind generated electricity. A collection system of power transmission lines to a centrally located substation for transmission to already existing national grids is required.

KEY CONCEPTS AND TERMS

Flux

Generator step-up (GSU)

Phase

Point of common coupling (PCC)

Slip

Step up

Synchronous

Vector

CHAPTER ASSESSMENT: WIND TURBINE TECHNOLOGY AND DESIGN CONCEPTS, PART 2

1. The principal function of the towerhead _____ is to convert the rotational mechanical power of the wind turbine to electricity.

2. The most cost effective and most frequently used generator in commercial wind farms is the:
   - ❏ A. variable speed system.
   - ❏ B. synchronous generator.
   - ❏ C. inductive generator.
   - ❏ D. asynchronous generator.

3. A power grid comprises high-voltage commercial power lines and is subject to government regulation.
   - ❏ A. False
   - ❏ B. True

4. The wind tower not only supports the towerhead, but it also holds the electrical:
   - ❏ A. microwave signal.
   - ❏ B. cable tray.
   - ❏ C. generator.
   - ❏ D. converter.

5. The wind tower GSU is a:
   - ❏ **A.** step-up transformer.
   - ❏ **B.** step-down transformer.
   - ❏ **C.** subtransformer.
   - ❏ **D.** special generator.

6. Because of the resistance losses of the low-voltage towerhead cable, the cable run needs to be as short as possible, and a GSU can be anywhere just so long as it resides on the wind farm.
   - ❏ **A.** True
   - ❏ **B.** False

7. Every US state has its own commercial high-voltage power grid.
   - ❏ **A.** True
   - ❏ **B.** False

8. A squirrel-cage generator is a:
   - ❏ **A.** synchronous generator.
   - ❏ **B.** variable speed generator.
   - ❏ **C.** induction cage rotor generator.
   - ❏ **D.** induction wound rotor generator.

9. Studies are starting to show wind farms have climate effects on the immediate region downwind of their presence.
   - ❏ **A.** True
   - ❏ **B.** False

10. A variable speed system can use both _____ and _____ generators.

11. A synchronous generator operates at what speed to match that of the grid frequency?
   - ❏ **A.** 60 Hz
   - ❏ **B.** Fixed speed
   - ❏ **C.** Speed higher then the grid to allow for line losses
   - ❏ **D.** Speed lower then the grid as the GSU will step-up the voltage before reaching the grid

12. The GSU will increase the tower low-voltage of 400 to 690 volts to a medium-voltage of somewhere between _____ and _____ for transmission to the wind farm substation via a collection system.

Design Factors Affecting Weight and Costs

THE TOWERHEAD OF A WIND TOWER holds a massive amount of weight—20 to 40 tons on average. Add to this sway and torque forces and other extreme conditions, and the core issue of any wind tower becomes its sheer weight. A towerhead that is heavier will generally cost more than a lighter one. Weight and cost of a wind tower are very closely related.

Wind farm development hinges on attaining the maximum amount of electrical production with the least amount of cost. Most wind energy designers concentrate their efforts on increasing electrical output and decreasing costs. There are several elements contributing to the design factors affecting wind tower weight and cost-return comparisons that will be covered in this chapter:

- Turbine component design and manufacturing
- Infrastructure of the wind farm, including grid integration
- Start-up operation and maintenance (O&M) costs
- Meteorology, wind farm landscape, and wind availability
- Lifetime operational expectancy

Offshore wind farms by their size and locale are vastly more formidable than land-based wind farms. The design of an offshore farm has several different patterns of cost and availability criteria. Also, operation and maintenance costs offshore require different interaction with the towerhead equipment. This chapter will focus on land-developed wind farms and the wind turbine at the towerhead. The towerhead itself accounts for 75 percent of the total developmental cost because the operational loads reside there.

Wind turbine designers use theoretical background with design drivers for specific analysis of the operation loads. They do so using quantitative analysis, formulated comparative studies, and design modeling. This chapter will cover the use of load-modeling practices for specific development of a wind turbine design.

Chapter Topics

This chapter covers the following topics and concepts:

- Identification and discussion of several different wind turbine load types
- The aerodynamic design of rotors
- Comparative cost analysis of rotor assembly power, energy capture, and machine costs
- Different main drive-train configurations, with their comparative weights and costs

Chapter Goals

When you complete this chapter, you will be able to:

- Discuss how different load types affect the operation of a wind turbine
- Relate the main aerodynamic designs of a wind turbine
- Compare the wind turbine power level to energy capture and related costs
- Convey how the different towerhead main drive-train configurations relate to weight and cost factors

Reducing the Cost of Electricity Production

Design factors of weight and cost do not contribute individually to turbine design. Reducing the cost of electrical production of these interrelating design features is a complex matter. If one component is redesigned, then you need to revisit all aspects of total design layout for cost comparatives. After each revision to a component's design, you may need to adjust an adjacent part or completely revamp a major feature.

A quantitative analysis is the first step for design choices. This gives you a good understanding of and baseline for each aspect of the design. It is a comparative approach that will also provide you with weight and cost benefits.

Three general design considerations are:

- Weight—A towerhead that is overall heavier (in both its structure and components) has proved to be generally more costly than a lighter one.
- Complexity—Components that are more straightforward in design and less complex, both in their process and in the overall structural size and design, are simply less costly.
- Materials—Standard materials (steel, copper, glass fiber) offer better cost reward than exotic composite materials (carbon fiber, rare earth magnets); however, the composite materials can offset and offer a cost reward due to less weight.

The actual components of these three factors can further be broken down into such design choices as rotor diameter, hub height, blade design, fixed speed versus variable speed, long versus short main drive train, or mechanical brake before gearbox versus after, as well as inductive versus synchronous generator, with or without the optional power converter control. With these added variables, you can begin to see the complexity of designing a wind tower with the best cost rewards.

Operational Loads of Wind Turbines

The operational loads of a wind turbine not only are highly complex but also are a collection of interactive loads that can be either fluctuating or quasi-static (apparently not in motion). In many instances they can function in both realms, one being no more or less critical when superimposed on the other.

Rotor blades are most at risk of load damage due to the nature of their operation of rotation and exposure to wind thrust and wind shear. The towerhead main drive train is also an at-risk area because it, too, has a rotational operation, it is linked with the rotor blades, and it's susceptible to bending forces.

Designers take special care to analyze in detail the load dynamics of these towerhead components and their adjacent counterparts. This will ensure dimensions are set to values within the dynamic operation levels that give the least amount of stress as possible.

Load Types

The operational loads of a wind tower all reside in the towerhead. There are three main load types and a number of sub-loads that must be considered when designing a towerhead. The three main load types can be classified as:

- Gravitational—The force of the Earth can be treated as an operational load when applied to the bending of the main drive train due to the weight of the rotor blades. This gravitational or mass/**inertia** type load can also be applied to the compression and tension in each load cycle, such as the **bending moment**, which is the rotational force of each rotor blade. Their self-weight initiates this process and also leads to tower tilt. Tower tilt is not an operational load but is especially prevalent as the towerhead yaws, changing the towerhead gravitational weight load on the tower itself.

- Aerodynamic—Thrust from the wind on the rotor blades is responsible for the most prevalent load force at the towerhead. Wind thrust, which is a perpendicular force to the plane of the blade, will carry on through from the blades to the hub and onto the main drive train. It will also create a torque force on the tower for a bending moment at the base of the tower; not an operational load but one of great importance if the tower is meant to reflect such forces and remain standing.

- Extraction—Load forces occur at every component in the towerhead where energy or power conversion takes place. For instance, the rotor blades convert wind energy to mechanical power via the collection of wind for rotational motion. This is an extraction load force. Loads derived from the extraction of energy, such as the torque forces on the main drive train, also are considered extraction load forces. The deviant forces of vibration and their attachment to the nacelle and tower load-path are a consequence of this type of load.

Most load forces are symmetrical near an axis such as the rotor blade pitch around its longitudinal axis or a torque force associated with the main drive train. Others are more asymmetric loads, for instance weight and wind shear, which produce a misaligned thrust but still a load force.

Fluctuating loads can either be random (stochastic) or deterministic. Random loads, such as those generated by wind gusts, are aerodynamic fluctuations over a period of time at indefinite intervals. Cyclic loads are deterministic, for instance when the self-weight of the rotor blades produce cyclic bending moments at the hub or when the rotor blade wind shear effect occurs on the main drive train.

It is cumulative fluctuations that result in fatigue damage. This damage can multiply by the amplitude of the fluctuation devaluing the designated intent of the load. The number of cyclic times it does so will cause a cumulative fluctuation that progressively reduces material strength.

Effect on Stress Levels

Most designers consider components that are productive up to 95 percent of their projected design lifetime a success. Measurable stress at the beginning of the intended lifetime, if considerable, will reduce the components' lifetime and will keep them from that 95 percent productivity goal. All load forces can cause stress on a towerhead's components. Once damage is initiated and exceeds the three main load types, it will produce stress levels that will cause cracking and component or material failure.

Avoiding Cracking

Cracking happens with the accumulation of damaging loads. Cracking is the physical outward sign that unseen load forces have exceeded their level of critical operational design. Once cracking occurs there is an acceleration of degradation. The rotor blade is a primary structure and the key component for which many stress and fatigue problems are most likely to occur.

Blades should be inspected before installation to detect manufacturing, shipping, or construction issues. This inspection takes place on the ground while the blade is staged next to the tower where it can be easily accessed for testing FIGURE 4-1. A technician will look for manufacturing defects inside and out, and also for shipping damage. This inspection is crucial to avoid future fatigue cracking from defective materials of construction.

Several techniques are used for inspection of load-bearing components. Three primary techniques are:

- Visual inspection, where technicians depend on their eyes, binoculars (after installation), and high-resolution cameras for detecting visible flaws.
- Tap testing and probing, whereby a technician taps the blade before installation and listens for variances in sound response, which can indicate a structural weakness or inconstancy of material. One such example would be an air bubble in the skin material, which might lead to debonding of laminated materials under load conditions.
- Thermographic camera to detect heat variances (stress creates heat) in the rotor blade laminations or other component resonant frequencies.

FIGURE 4-1 Wind farm technician inspecting blades staged for installation on a wind tower.
© iStockphoto/Thinkstock

Avoiding Failure

The two most critical means to avoid component failure are to avoid exceeding the ultimate tensile strength of the component material and to operate within the cyclic load design (number of cycles necessary to initiate damage). The ratio of maximum applied cyclic stress to a material's tensile strength in conjunction with the designed lifetime gives you the level of damage to expect where failure occurs.

Rotor blades are subjected to cyclic forces no matter how much design influence is taken into consideration. Designers take great pains not to exacerbate the effect of load forces. Connecting joints, sharp changes in blade profile, and hub junction are all stress-related points that designers consider to avoid failure. Choice of blade material is also a major consideration in avoiding failure at the towerhead as is exceeding the tensile strength once a material is chosen.

Design Considerations

Before a proposed design is permanently incorporated into a wind turbine system it passes through modeling tools and methodologies to predict its long-term stress, fatigue, and functionality. However, in advance of the modeling process there is the question of what design features should be addressed and considered in the first place **FIGURE 4-2** . These design considerations can be categorized into three areas that affect design weight and cost of wind turbine component choices:

- Design drivers—Address design features over a large range of turbine sizes and different component configurations. Design drivers attempt to imitate the proposed design but in a much simplified manner. They are a means to minimize spending unwarranted time addressing complicated issues before determining the viability of a design.
  A stationary load is used first to establish operational set points before a fluctuating real case load is addressed. Several current design drivers for the wind turbine industry are low-mass nacelle arrangements, increase in rotor diameter, increase rotor tip speed, greater bandwidth for variable speed, higher efficiency in low-wind conditions, lighter materials of construction, and overall improvements in aerodynamic performance. Designers apply design drivers for operational load calculations at critical intervals and/or vulnerable parts of the towerhead system.
- Service factors—Internal mechanical resonant loads are directly related and react directly to the affect dynamic loads have on them. They act as an interface to the applied loads such as power control at the blades, rotor rotation at the hub, and main drive-train rotation. Internal load response to applied loads appears as a physical distortion in the materials of construction.
- Structural response—A component's natural and forcing frequencies, if left unchecked, will compound, building up large deformation and

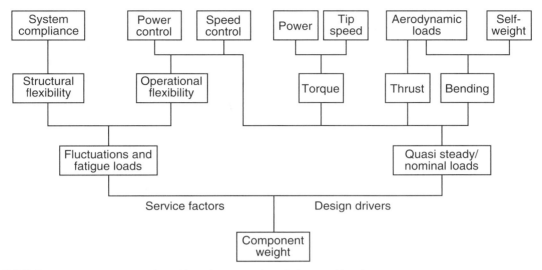

FIGURE 4-2 Tree diagram showing a hierarchy approach to design considerations.
Adapted from Harrison, R., Hau, E., & Snel, H. (2000). *Large Wind Turbines Design and Economics*. West Sussex, England: John Wiley & Sons Ltd

vibration damage. A stiff component system is less likely to withstand wind fluctuations and suffer the highest incidents of structural fatigue. A soft system, which responds more rapidly will be more compliant and show less signs of fatigue (such as the flexibility in rotor blades).

Aerodynamic Design of Rotors

Aerodynamic optimization of the rotor assembly is foremost on the manufacturer's and designer's mind. This is the initial and only point of contact with the wind energy source, and establishes all downstream component configurations whether for a new design or adaptation of an existing design.

The designer has two sets of primary design parameters to consider—operational and geometrical. Operational parameters establish a sequence of operations. This sequence of operations establishes the physical requirements needed to fulfill the design criteria. Geometric parameters enter into the design process after the operational parameters are set.

For comparison, the operational parameters of the rotor assembly are:

- Speed control—Fixed versus variable rotor assembly for speed of operation
- Blade pitch control—Fixed versus variable pitch angle (angle of attack) of rotor blade for different power coefficient (C_p) to tip-speed ratio curves
- Power control—Pitch control, stall control, and active yaw control for power limitation

Geometrical Parameters

Geometrical parameters are structural, having physical presence and dimension. Rated electrical power is a common starting point for determining geometrical parameters. It dictates the rotor tip-speed ratio that leads to rotor diameter required and so on down the line to the other geometrical parameters to meet the requirement frequency (60 Hz) of the power grid.

It is difficult for geometric parameters to mimic modeling tools that use design drivers for a large range of sizes. The designer must make geometric choices, settle on criteria, and move forward within the set parameters. There are four distinct geometric parameters of the rotor blade design, each having aerodynamic influence on the rotor assembly:

- Number of blades—This must be determined before all other aerodynamic decisions are considered. This choice sets the precedence of all other design decisions.
- Aerofoil section—At a specific blade radius, the aerodynamics of that particular section of the blade are determined, most importantly to establish the blade chord to blade thickness ratio.
- Profile—The blade chord tapers off (or decreases) in dimension from blade root to blade tip.
- Twist angle—For each blade radius dimension, twist angle will be set to a specific degree-of-angle along the trailing edge of the blade.

Design criteria already established by the manufacturer's practicalities (factory tooling already in place), structural design, and transportation limitations figure heavily into design parameter choices wherever possible and convenient.

Coefficient of Rotor Performance

The coefficient of rotor performance, most often referred to as the coefficient of power or power coefficient (C_p), establishes a means to rate a turbine's percentage of wind extraction. The power coefficient represents the ratio between the potential wind energy to mechanical power extracted.

> **NOTE**
>
> For a list of the symbols referred to in this chapter, see the Glossary of Symbols in the Appendix.

A designer should consider all four of the geometrical parameters listed above when determining a rotor assembly power coefficient. The higher the power coefficient of a rotor assembly, the higher is its design efficiency for extraction of wind energy from the airstream. A power coefficient of 0.45 to 0.50 is considered very high for a horizontal wind turbine.

Choice of Aerofoil Section, Blade Chord Profile, and Blade Twist Angle

Aerofoil section, blade chord profile, and blade twist angle are the determining factors for how high a tip-speed ratio (λ) can be anticipated. However, a high tip-speed ratio does not necessarily mean a higher power coefficient. **FIGURE 4-3**

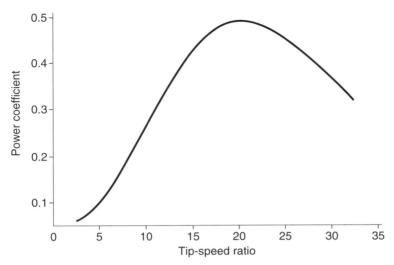

FIGURE 4-3 Power coefficient to tip-speed ratio of stall-controlled three-blade rotor.

Data from Harrison, R., Hau, E., & Snel, H. (2000). *Large Wind Turbines Design and Economics*. West Sussex, England: John Wiley & Sons Ltd

illustrates a typical performance curve for a stall-controlled, three-blade rotor assembly where tip-speed ratio is plotted against the power coefficient. Note how at higher speeds the power coefficient begins to drop off.

The low tip-speed ratio denotes low wind speeds; at the most aerodynamic efficiency of the blade, the curve peaks, but it begins to drop again at the higher speeds. That's because this is a stall-controlled rotor assembly and is designed to avoid the higher wind speeds. The operational parameters you choose will deter-mine the aerodynamic curve you'll work toward when you plot your power coef-ficient against your tip-speed ratio.

The designer will work to influence the performance curve of a blade design by the lift-to-drag forces at each aerofoil section of the blade radius. The blade chord length and twist angle both will also figure into this lift-to-drag force.

Depending on the operational parameters, the performance curve is the designer's tool to matching the geometric parameters to give the broadest speed tip-speed ratio curve at the maximum power coefficient. For example, designers seek a curve that goes high and stays there over the widest spread of tip-speed ratio. This is accomplished by using a different pitch angle on the rotor blades for varying wind conditions, which allows for a longer sustained maximum curve of the power coefficient in low to moderate winds **FIGURE 4-4**.

Determining Optimum Rotor Tip Speed and Number of Blades

Optimum rotor tip speed relates directly to the number of rotor blades via their respective maximum achievable power coefficient. The optimum power coefficient

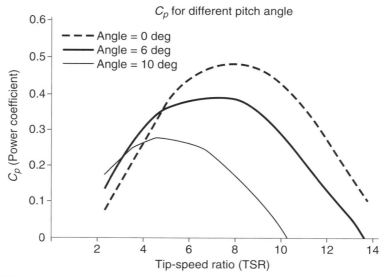

FIGURE 4–4 C_p for three different pitch angles.
Data from National Renewable Energy Laboratory

for any rotor assembly is called the Betz limit, with a power coefficient of 0.59. There are three aerodynamic loss mechanisms that contribute to the maximum power coefficient being lower than optimum:

- Blade profile drag due to each designated blade aerofoil design section's lift-to-drag ratio with drag increasing the higher the tip-speed ratio **FIGURE 4–5**.
- Tip loss effect due to the non-uniformity of the induced velocities of the trailing blade tip vortices in the wake of the blade. This loss mechanism decreases with increasing number of blades and increases with tip-speed ratio.

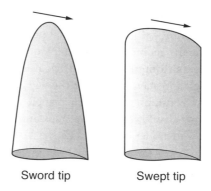

Sword tip   Swept tip

FIGURE 4–5 Blade tip geometries demonstrating trailing vortices.

◾ Swirl in the rotor blade wake due to tangent vortices coming off the trailing edge of the blade acting on the airflow. Swirl losses will be smaller at higher tip-speed ratios.

Taking all three loss mechanisms into consideration, a three-blade rotor can obtain a maximum power coefficient of 0.5 with a design tip-speed ratio of approximately 7.5. A two-blade rotor must hold a maximum tip-speed ratio of between 8.0 and 10.0 in order to achieve a power coefficient much less than a three-blade rotor. Hence, the three-blade rotor is used more often, and it's the conventional choice of most large commercial wind turbines.

Also the two-blade rotor has a twofold situation whereby it is more sensitive to blade environmental degrading (soiling) due to its higher tip-speed ratio and it also has the added issue of high frequency noise possibility. A single-blade rotor will have a tip-speed ratio in the region of 14 to 15, with a higher soiling and noise possibility than a two-blade rotor.

Solidity

Rotor blade solidity (S) is the ratio of the total blade surface facing the air stream to the full rotor assembly area that the rotor blade passes through. By reducing solidity of the rotor blade, optimization of the rotor speed is accomplished. Solidity is inversely proportional to tip-speed ratio: As solidity decreases, tip-speed ratio increases **FIGURE 4-6**.

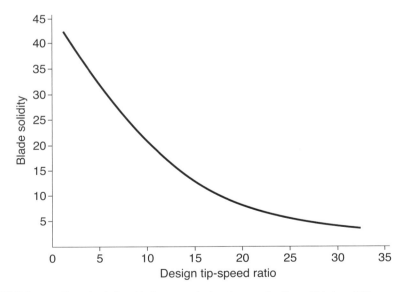

FIGURE 4-6 Plotted relationship between design tip-speed ratio and blade solidity.

Data from Harrison, R., Hau, E., & Snel, H. (2000). *Large Wind Turbines Design and Economics.* West Sussex, England: John Wiley & Sons Ltd

A rotor blade comprises two very distinct and independent components: the load bearing spar (leading edge structure) and an aerofoil form (fairing). Solidity relates to modeling cost of the rotor blade as the determining factor for blade fairing costs.

Aerofoil

Wind turbine design traditionally has made use of general aviation aerofoil profiles. With the advent of growth in size of wind turbines, this is changing. Wind turbine aerofoil profiles are now becoming more specialized and criteria specific to the wind energy industry are becoming more prevalent.

Specific criteria for specialized wind turbine aerofoil profiles are:

- Lift-to-drag ratio increases as megawatt-size turbines are being manufactured and put into energy production.
- Operation of towers under soiled conditions is far more prevalent as many wind farms are being placed in desert environments. Environments without much rain to help clean rotor blades but with high insect incidence have higher O&M costs.
- Aerofoil thickness is especially critical at the rotor blade root. General aviation profiles design at a thickness ratio of no more than 21 percent, while a rotor blade thickness is at a 25–40 percent ratio, while also lowering the blade mass.
- Stall control of the rotor assembly requires the aerofoil profile to meet criteria that allow the lift coefficient not to drop off too fast after a stall is initiated, while the drag coefficient rises exponentially.

Vibration Issues

Aerodynamic forces of wind speed and mass effects (gravitational forces) not only contribute to excitation of the rotor assembly but also produce spring-like and **damping** actions. Vibration issues caused by excitation are of these types:

- Deterministic aerodynamic—Forces are caused by averaged anticipated conditions, such as both vertical and horizontal wind shear. As discussed earlier, when the rotor blade passes the wind tower the inevitable wind shear between the tower and the blade is a constant and always expected.
- Stochastic aerodynamic—Forces are turbulent deviations felt by the rotating rotor blades that are not the norm, such as wind-speed fluctuations. These forces are transferred to adjacent components and support structure giving rise to frequency shifts at random intervals. Also bending moments across the blade swept area between the blade tip and root add up to a substantial increase in resonant frequencies. These forces not only add to the fatigue of the blade structure but all the way downstream to the turbine hub, bearings, main drive shaft, and support tower load path.

- Gravitational—The force of the rotor assembly center of gravity, in theory, will rotate exactly on the horizontal axis of the wind turbine. In reality it never reaches this precision, and the offset causes gravitational and centrifugal fluctuations, setting up bending moments on the main drive train radiating out to all other adjacent components as vibration. It will also experience torque fluctuations from the **coriolis force**, creating a resonant frequency.

Natural frequencies inherent in the rotor blade will occur due to flap and lead-lag dynamics depending on the stiffness of the blade. As blade size and length increases with the current trend toward megawatt turbines, the overall strength of the blade needs to increase as well. Higher levels of centrifugal and cyclic gravitational loads require what is called a *stiff blade* with even higher blade frequencies.

Blade Weight

The design of the blade, with its resultant weight, represents a substantial expense to the overall cost of wind turbine production. Each extra measure of blade length requires extra strength, adding to the structural weight. Materials of construction are also a major consideration for weight reduction, as are conceptual influences of blade flexibility, control (pitch or stall), and operational speed (fixed or variable).

Several other factors that determine the weight of a blade's load-carrying structure are:

- Rotor diameter (D)—The most important factor for large-scale commercial wind turbines. The rotor blade self-weight, as proven via modeling tools, is the main design driver over aerodynamic forces. Also rotor diameter dictates the tower hub height, adding considerably to total tower cost factors.

- Design tip-speed ratio (λ_d)—Contribution to total blade mass is relatively small but the factors determining this ratio are not. The blade spar mass is determined by the number of blades whereby two blades have a higher tip-speed ratio than a three blade. Also wall thickness of the blade and blade cord dimension are determinant upon the number of blade selection.

- Design wind speed (V_d)—Used as a representative of the site mean wind speed (rated wind speed ($V_{d,} V_R$) for determination of fatigue damage for a wide range of wind speeds.

- Relative profile thickness (t)—From blade root to blade tip, it is critically important to weight and cost. The influence of the relative profile thickness to chord ratio is a standard value of 21 percent with up to 40 percent for thick aerofoil profiles. The blade root profile thickness is the most critical.

Rotor Power Level, Energy Capture, and Machine Costs

Rotor power level, energy capture, and machine costs of the wind turbine are determined by the three fundamental operating parameters mentioned earlier in the chapter: speed control, blade pitch control, and power control. Designers use these variables as foundation choices for creating a model of a wind turbine.

These three fundamental operating parameters for design consideration of optimizing rotor power and energy capture are:

- Blade pitch control—Measures are used to limit a variable speed rotor to operate at an optimum tip-speed ratio (λ_d) thereby allowing the maximum power coefficient (C_{Pmax}) over as much wind speed range as the system will permit. For example, the rotor speed changes with the wind speed due to rotor blade angle-of-attack variances (see Figure 4-4).

- Power control—Strategies for limiting power are required to avoid excessive overloading of the wind turbine and its rated power. This is accomplished by pitching the blades as mentioned above, but only until such time as the rotor begins to slow down (feather) or stall. Power control is also accomplished with a passive stall control via the rotor blade that is fixed and designed so it goes into a stall at a specified high wind speeds due to the blade design dynamics. A complete cut out of power generation happens at extremely high wind speeds of short duration.

- Speed control—Consideration starts with the designer choosing a fixed rotational speed of the rotor assembly (Ω) for a predetermined rotor diameter (D) thereby giving a fixed rotor tip speed (V_T). The purpose of fixing the rotor tip speed is to formulate the tip-speed ratios for all site wind speeds. Next it is plotted against the power coefficient (C_p) for determination of a power coefficient curve (see Figure 4-3).

$$V_T = \frac{\Omega D}{2}$$

Define Wind Speed Formula

Each wind turbine is given a rated power output (kW) that is best understood via the wind speed formula. **Rated power** is the generated power output of a wind turbine in kilowatts. This equation derives the maximum rated power (P_R) of a turbine by defining the maximum power coefficient (C_p) that can be realized for a specified wind speed range.

The rated power (P_R) of the rotor assembly at a particular wind speed can be formulated as follows:

$$P_R = \frac{1}{2} \rho_a C_P V^3 \pi \left[\frac{D^2}{2} \right]$$

On a fixed-speed wind turbine the design wind speed (V_d) is defined as the wind speed where the power coefficient is at its maximum. This is derived by the plotting of the design tip-speed ratio (λ_d) to its corresponding power coefficient (see Figure 4-3) for a curve showing a maximum power coefficient (C_{Pmax}) such that when $V = V_d$ then $\lambda = \lambda_d$ and $C_p = C_{Pmax}$.

Once the value of the design tip-speed ratio (λ_d) is set by the designer, this allows the selection of a design wind speed (V_d) for locking in the rotor tip speed (V_T). The formula looks like this:

$$V_T = \lambda_d V_d$$

Variable Wind Speed

On a variable wind speed turbine, the design wind speed cannot be formulated as for a constant (fixed) wind speed turbine (as shown in the formula above). On a variable wind speed, the rotor assembly speed adjusts with the wind speed to maximize the power output.

A variable wind speed turbine is allowed to function over a wide range of wind speeds (see Figure 4-4) while still operating at the design tip-speed ratio (λ_d). The only time this control strategy does not comply is when the output power is exceeded, as when the cutout speed is reached. At this point the turbine is taken off the grid and parked.

Generated Power Dependencies

Public electric utilities, in order to allow a wind farm to connect to their power grid, require wind turbines to generate a high capacity of power. The utilities want a turbine's capacity to have a consistent high average power output, or megawatt (MW), in relation to the turbine's maximum power capabilities. For example, one turbine may be rated at 1.5 MW of power at 30 mph wind speed, while another turbine also rated at 1.5 MW may not make its rated power output until it gets a 40 mph wind speed. This variance constitutes some analyses for determination of an annual average of generated power.

The megawatts produced by a wind turbine connected to a national power grid are evaluated annually by examining the power curve of the wind turbine along with the range of wind speed probability distribution. This is a two-part process represented by two separate formulated evaluations:

- Rated power curve (generated power)
- Wind speed probability distribution

Rated Power Curve

The first part of the evaluation represented by the rated power curve is expressed by the following equation:

$$P = \eta_t \eta_g \frac{1}{2} \rho_a C_p V^3 \pi \frac{D^2}{2}$$

Three key concepts make up this power curve equation and can be looked upon as three individual dynamics:

- **Aerodynamic rotor power coefficient** $C_p(\lambda)$—Expressed in the previous equation as the aerodynamic power/unit area ($\frac{1}{2}\,\rho_a\,C_p\,V^3$). This calculates the power available in the wind, which is determined by the air density ($\frac{1}{2}\,\rho_a$) and wind speed (V^3) along with the efficiency of the wind turbine power coefficient (C_p) or expression of the wind turbine efficiency of power conversion.
- **Gearbox efficiency (η_t) and generator efficiency (η_g)**—A mechanical apparatus is never 100 percent efficient and neither are the wind turbine gearbox and generator. In their functions of converting the energy of a rotating shaft into electricity, there are losses due to friction from bearings and the gearing. There are also magnetic drag and electrical resistance losses in the generator. These efficiency percentages are high over most of the power curve but can drop off sharply at levels below 30 percent of maximum rated power P_R.
- **Swept area ($\pi[D/2]^2$)**—The area available for capture of the wind as scribed by the rotor blades is most critical. As discussed previously in the chapter, the larger the swept area (in square feet) the greater the extraction of the wind is possible.

These three parameters are based on the mean wind speed taken from meteorological sites previous to the installation of a wind farm and from the design drivers the designer used in the development phase of the turbine. With these three parameters, the designer now estimates, projects, and tweaks the power that is produced by a wind turbine installed at a selected location **FIGURE 4–7**.

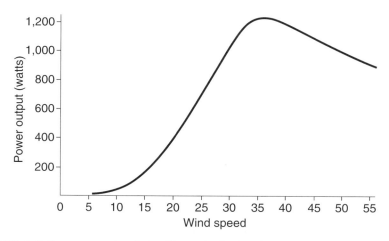

FIGURE 4–7 Power curve of a wind turbine projecting how large the electrical output will be for different wind speeds.

Data from OtherPower.com

Annual Energy Output Formula

The second part of the public electrical utility evaluation is the wind speed probability distribution requirement. This is an evaluation of the specific wind site in question and how many hours per year and at what wind speeds the site in all probability will experience. On an annual basis, the public utility requires a high ratio of average energy output to maximum tower rated capacity output. This is a calculated percentage called the capacity factor.

This wind speed probability requirement is best approximated by a Weibull distribution $\Phi(V)$, a theoretical probability curve taking into consideration a good approximation of relatively common wind speed. In **FIGURE 4-8**, the horizontal axis shows wind speed, and frequency is listed on the vertical axis.

The annual energy output in kWh (E_y) can best be expressed mathematically as such:

$$E_y = 8760 \int_{V_{in}}^{V_{out}} \eta_t \eta_g P_S(V)\Phi(V)dV$$

Wind passing through the rotor swept area will lose speed as energy is extracted. The annual energy equation above is defined by the **integral (∫)**, which is the interval of numbers between two set points. The first is the wind speed before the swept area (V_{in}) and the second is the wind speed after passing through the swept area (V_{out}). This is expressed as a product of generated

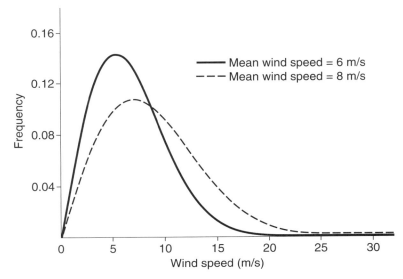

FIGURE 4-8 Chart depicting the Weibull distribution.
Data from Canadian Wind Energy Atlas

electrical power and the frequency of occurrence of the different wind speeds formulated as:

- Gearbox and generator efficiency: $\eta_t \, \eta_g$
- Mechanical shaft power: $P_s(V)$
- Weibull distribution: $\Phi(V)$
- Design wind speed: V_d

The integral (kW) is adjusted to meet an annual figure by multiplying the number of hours in a year (8,760 hours). With the evaluation of the expression for annual energy output, you can now use the influence of these parameters to investigate and model the total annual energy output of a wind turbine.

The public electrical utilities prefer the expression of this process of evaluation in non-dimensional form known as the capacity factor (C_f) or *load factor* of the wind turbine. This formulated quantity can be expressed as:

$$C_f = \frac{E_y}{8,760 P_R}$$

The resultant capacity factor of the above formula is a non-dimensional number the public electrical unities use for measuring the productivity of a wind turbine. This number compares the energy output of the turbine based on rated power (P_R) or maximum installed power of the wind turbine if it had run at full capacity all year, to the actual average produced power (E_y). This is a ratio number and usually given in a percentage. The electrical utility companies are looking for a percentage of 25 percent to 40 percent.

Maximum Energy Capture Formula

Capacity factors of a wind turbine generally are too low (20 percent range) to be accepted by the electrical utility companies. The maximum energy capture ratios of design wind speed (V_d) to the annual mean wind speed (V_m) can be formulated and used to better the capacity factor. This is done in order to meet the 25 percent to 40 percent capacity factor required by the electrical utility companies.

It should also be noted that lower values of the rated wind speed (V_R), which also reflect rated power output and higher values of design wind speed (V_d), will lower the unit costs of electricity. Hence, there is a fine balance between the lowering and raising of design wind speed to rated wind speed while still attempting to meet the electrical utility requirements.

Designers have formulated a matrix in which to do comparative analysis of the effect wind speed variances have on optimizing annual output to cost. The maximum energy capture of a wind turbine can best be expressed by use of the non-dimensional X factor whereby X_d represents design wind speed and X_R represents rated wind speed. By formulated ratio comparison, the designer can optimize a turbine's wind capture capability.

TABLE 4-1 VALUES OF X_d AND X_R FOR THREE DIFFERENT CONTROL STRATEGIES SHOWING MAXIMUM ENERGY OUTPUT		
Control strategy	X_d	X_R
Fixed speed pitch	1.35	2
Fixed speed stall	1.35	(2)
Variable speed	-	2

Data from Harrison, R., Hau, E., & Snel, H. (2000). *Large Wind Turbines Design and Economics.* West Sussex, England: John Wiley & Sons Ltd

Maximum energy capture can be depicted by the following two ratio comparisons:

$$X_d = V_d / V_m \text{ and } X_R = V_R / V_m$$

where V_m is the annual mean wind speed at the wind tower site. Resultant values of X_d and X_R are tabulated for the three different control strategies you intend to use for your model **TABLE 4-1**. By varying the design and rated wind speed of a model, various optimization factors can be realized and manipulated as a starting point for design choices.

The qualitative effects of the independent variables X_d and X_R are:

■ Increase in X_d results in an increased tip-speed ratio and rotational speed. With a constant X_R, the nominal drive-train torque decreases. If X_d is larger than 1.35 (see Table 4-1), the maximum energy output decreases.
■ Increase in X_R results in a higher installed generator power. For values lower than 2, energy production increases but turbine costs may increase. With a constant X_d, the capacity factor C_f falls. As discussed above, the power grid electrical utility companies require a capacity factor above 0.25 so this relationship requires watching.

Now you have a method for comparative analysis by which you can compare design choices against optimal power output and cost of construction to make quantitative fact-based design decisions.

Drive-Train Configuration

Another design driver to consider is the drive-train configuration, which can help reduce weight and costs. The main drive-train layout not only determines the weight of the support bedplate structure but a shorter shaft can result in a weight reduction. Also, a short shaft reduces shaft-bending moments, allowing for a 25 percent decrease in shaft diameter, to reduce weight by 50 percent.

Weight Factor

The overall length of the main drive train depends upon the style of layout and components coupled to its shaft. A fixed speed direct shaft does not require a gearbox, for instance, while some layouts allow for a shorter length of shaft. This is a direct weight reduction due to a shorter main drive train and relates to significant resonant frequency reduction as well. Rotor lag motion, blade passing wind shear, and the inherent rotation of the main drive train itself add to the resonant frequencies of the drive shaft. Shorter drive trains experience less resonance frequency.

Bedplate

As a primary function, the nacelle bedplate acts as a carriage support for the rotor assembly, the main drive train, and generator. A secondary function is its ability to act as an essential towerhead load path receiving resonant frequencies (rotor load thrust, torque, vibrations) from the components it supports and transferring these frequencies to the tower load path. It is also designed to uphold the bending moments produced by these loads and hence a short shaft will give less bending moments than a long shaft with an overall weight reduction and cost reward.

WORLD WIND ENERGY STATISTICS

The wind energy industry is increasing exponentially every year in capacity of production and the percentage of deliverable electricity **FIGURE 4-9**. The expected total wind capacity is forecast to exceed 200 GW and equaling over 2 percent of the global electricity consumption. This wind sector for the first time is expected to offer over one million jobs and is projected to double wind capacity every three years.

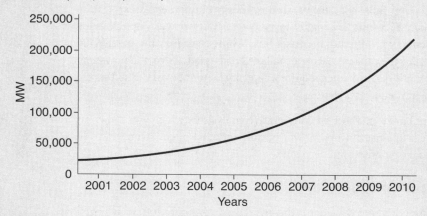

FIGURE 4-9 The increase of total world capacity of wind-generated electricity.
Data from World Wind Energy Association WWEA 2010

WORLD WIND ENERGY STATISTICS (Continued)

The world finance sector now recognizes wind energy as a viable area for investment—especially as deregulation is becoming more prominent. Both North America and Africa now have laws for guaranteed grid access similar to the US deregulation in 1978. These laws provide for long-term contracts and purchase prices based on the cost of renewable energy generation. In addition to viability, wind energy is proving able to help stabilize overall energy prices while reducing dependence on electricity from fossil fuels and nuclear energy—which, for many countries, involves imports, often from unstable parts of the world.

Commercial wind energy is now used in 82 countries, with China and the United States the largest markets, accounting for 38.4 percent in 2009. Nine other countries with turbine sales in the 0.5 to 2.5 GW range are Spain, Germany, India, France, Italy, Britain, Canada, Portugal, and Sweden. Countries with a wind farm capacity of 100 MW or more number 35 worldwide. Offshore wind turbines are slated to be a future trend at 1.2 percent of the total worldwide wind capacity and a growth projected at 30 percent—just under the general growth rate of wind power.

CHAPTER SUMMARY

A wind tower looks so simple and sleek in its design, but it is not. A wind tower is a very complex structure developed and designed expressly to capture one of the most abundant energy sources—the wind.

Wind energy developers want the most cost-efficient towers producing the highest yields of power output possible. Designers strive to give the developers just that by means of very detailed design drivers. Once a turbine's design parameters are established, the real work begins. A designer must adjust, tweak, and formulate all the components and their force dynamics into a cohesive unit, producing not only a high rate of power, but a high percentage of annual output in order to comply with public utilities' requirements for connection.

KEY CONCEPTS AND TERMS

Bending moment
Coriolis force
Damping
Inertia
Integral ($\int$)
Rated power

CHAPTER ASSESSMENT: DESIGN FACTORS AFFECTING WEIGHT AND COSTS

1. The turbine accounts for 75 percent of the wind tower's:
   ❏ A. structural costs.
   ❏ B. geometrical loads.
   ❏ C. development costs.
   ❏ D. aerodynamic loads.

2. Three areas for weight and cost comparisons are: (Select three.)
   ❏ A. materials.
   ❏ B. complexity.
   ❏ C. components.
   ❏ D. weight.

3. The operational loads of a wind tower all reside in the towerhead.
   ❏ A. True
   ❏ B. False

4. Gravitational force of the Earth can be treated as an operational load when applied to:
   - ❑ **A.** tower tilt.
   - ❑ **B.** weight of the rotor blades.
   - ❑ **C.** weight of the turbine generator.
   - ❑ **D.** bending moment of the main drive train.

5. What are the two types of fluctuating loads? (Select two.)
   - ❑ **A.** Aerodynamic
   - ❑ **B.** Stochastic
   - ❑ **C.** Cumulative
   - ❑ **D.** Deterministic

6. Three primary techniques used for inspection of load-bearing components are: (Select three.)
   - ❑ **A.** taping test.
   - ❑ **B.** visual inspection.
   - ❑ **C.** wind tunnel test.
   - ❑ **D.** thermographic camera.

7. Design _____ attempt to imitate the proposed design but in a much simplified manner.

8. The rotor blade has four geometrical parameters: numbers of blades, profile, twist angle, and _____.

9. The designer has two sets of primary design parameters: operational and:
   - ❑ **A.** aerodynamic.
   - ❑ **B.** geometric.
   - ❑ **C.** aerometric.
   - ❑ **D.** performance.

10. C_p is the nomenclature for which three choices below? (Select three.)
   - ❑ **A.** Power coefficient
   - ❑ **B.** Potential coefficient
   - ❑ **C.** Coefficient of power
   - ❑ **D.** Coefficient of rotor performance

11. What are the three aerodynamic loss mechanisms for a rotor blade? (Select three.)
   - ❑ **A.** Swirl in rotor blade wake
   - ❑ **B.** Tip loss effect
   - ❑ **C.** Number of rotor blades
   - ❑ **D.** Blade profile drag

12. Two turbines each rated at 1.5 MW of power will have the same annual average of generated power.
❏ **A.** True
❏ **B.** False

13. Wind speed probability is best approximated by a theoretical probability curve called the _____ distribution.

14. Wind passing through the rotor swept area will _____ as energy is extracted.
❏ **A.** increase pressure
❏ **B.** increase speed
❏ **C.** maintain speed
❏ **D.** lose speed

15. A mechanical apparatus is never 100 percent efficient, and the wind turbine gearbox and generator are no different. In their functions of converting the energy of a rotating shaft into electricity there are efficiency losses due to which three choices below? (Select three.)
❏ **A.** Friction
❏ **B.** Structural load
❏ **C.** Magnetic drag
❏ **D.** Electrical resistance

Determining Wind Turbine Weight and Costs

PREDICTING THE PERFORMANCE of a wind turbine while it is still in the design phase is tricky business. Modeling methodologies and tools are used to bridge the gap between theoretical formulations and the reality of a physically operating wind tower. The identified design drivers and service factors are now given specific parameter dimensions to be used in virtual experimentation for weight and cost analysis. Service factors are variable values independent of the load and weight formula but relevant to the calculated design decision.

Virtual experimentation is crucial not only to prototype design but for the improvement and updating of wind turbine designs already in production. The wind turbine designer uses scaling model tools reflecting various sizes of wind turbines to understand better and predict more accurately the complex physical interactions of various turbine loads.

Scaling models reflecting various sizes of wind turbines offer a means for observable measurement across a wide range of wind turbine operating conditions. Potential technology pathways and the barriers to size and subsequent weight increase can now be identified. Modeling of individual component load analysis to overall turbine performance involves load simulation, measurement, and comparison. This then will grant the designer the means to more accurately quantify and predict aerodynamic and structural responses at given load stations as well as predict component interfacing.

Researching and testing under simulated load predictions provides the assessment of inconsistencies and discrepancies by the designer. Virtual experimentation highlights flaws and limitations for refinement; yielding load validations before design implementation. Once such barrier and flaw limitations are identified, it is expected the designers will find ways around them.

Chapter Topics

This chapter covers the following topics and concepts:

- Performance prediction by means of modeling methodologies and tools
- The Sunderland model
- Calculating weights for cost modeling
- The steel tower: the most critical support structure in the wind development system

Chapter Goals

When you complete this chapter, you will be able to:

- Relate how the wind designer can accurately predict, quantify, and understand the complex physical interactions characterized by a wind turbine's performance
- Discuss the various modeling tools and methodologies for predicting turbine component respective weights
- Formulate and calculate the weight of various turbine components

Weight and Cost Models for Predicting Performance

Virtual modeling of a wind turbine allows the designer to mimic the load process in a simplified way as to establish a virtual representation of the turbine design processes and prototype. It can also allow the designer to take an established tower design and improve and supplement what has already been designed. To accurately predict, quantify, and understand the complexity of the physical and dynamic interaction of load characteristics, you must consider the weight relevance of key turbine components such as blade weight and load forces projected by these blades.

Weight of Blades

The actual physical weight of the rotor blades is mostly due to the composite materials of construction of the blade itself. Most important, once you determine the weight of the rotor blade, you'll use it to calculate other turbine components. This blade weight directly relates to the structural makeup of the pitch mechanism, the low-speed shaft, and the nacelle bedplate.

The gravitational weight load of the rotor blade will place bending moments on the horizontal drive shaft, most specifically the high torque, low speed main drive shaft immediately behind the rotor hub. This bending moment of the drive shaft will in turn place stress on the nacelle bedplate dictating its size and structure. These bending forces sustained during normal operating conditions also affect the aerodynamic forces of stresses.

It is also important to consider the pitch mechanism. The higher the physical blade weight, the more heavily built the pitch mechanism needs to be. Also of critical weight consideration is the aerodynamic structure (spar and aerofoil fairing) of the blade at each design blade radius **FIGURE 5-1**.

Aerodynamic Forces

Aerodynamic forces offering the least resistance to wind inflow are identified as bending forces sustained by the blades and the hub. These forces are used to estimate stresses incurred during normal and various running conditions.

Fatigue

Fatigue is difficult to incorporate in a weight and cost model, as it is inherent in the design process but not dimensional. Different levels of fatigue occurring in each case load must be taken into account and the approach of model comparisons made against calibrated cases.

Stiffness

You'll also need to consider the stiffness of components, such as the rotor blades and tower, throughout the modeling process as important resonant frequencies.

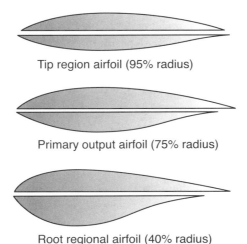

Tip region airfoil (95% radius)

Primary output airfoil (75% radius)

Root regional airfoil (40% radius)

FIGURE 5-1 Illustration of blade aerofoil at three different design blade radii.
Adapted from Office of Scientific and Technical Information, OSTI, U.S. Department of Energy

The design process works to diminish such forces for both the components and subsystems of the wind turbine.

The Sunderland Model

The Sunderland model was developed in the early 1990s in the United Kingdom at the University of Sunderland, in England. The initial model established the first set of scaling tools for utility-size wind turbines capable of commercial use in the class ranges for that time. (**Scaling** refers to size, or change in size, as in reference to scaling models reflecting various sizes of wind turbines.)

As commercial wind turbines grew bigger and more sophisticated, many scaling model attempts became outdated before they could be used. The commercial wind industry, however, began to stabilize around the late 1990s with the standard acceptance of a three-blade, upwind design, with modeling tools honing in on this particular design model.

An updated Sunderland model revisited the commercial market needs in the late 1990s, setting valuable scaling tools for predicting the impact of turbine size to its load components. In 1999, the US Department of Energy (DoE) began its Wind Partnerships for Advanced Component Technologies (WindPACT) project. This work determined the impact of increasing rotor diameter and turbine configuration on total cost of energy. The resultant conceptual designs focused on megawatt-size turbines establishing potential technology pathways in the hopes of leading to more cost-effective designs. When scaling tools of superior form were developed from the WindPACT studies, the Sunderland model was abandoned in the United States and new scaling tools were defined.

Rated Torque

When you are working with modeling tools, the first and single most important design driver of a turbine model in determining size and weight is rated torque (Q_R). This is the rotor torque at rated wind speed (V_R), which is directly related to the rotor diameter (D). It is also inversely proportional to rotor tip speed (V_T), as can be indicated by the following formula:

$$Q_R = \frac{1}{16} \rho_\alpha C_p \pi \frac{V_R^{\,3}}{V_T} D^3$$

Wind inflow extracted through the rotor assembly imparts a torque on the blades causing the rotation about the horizontal axis. Torque is what allows the rotor blades to apprehend energy from the wind. Inversely, thrust is a byproduct of torque as the blade's inherent gravitational, aerodynamic, and friction loads kick in to set up a resistance. Therefore, thrust (T_R) affects downstream components by placing weight on the main drive shaft bearings, the nacelle bedplate, the tower, and tower foundation.

Calculating Weights for Cost Modeling

To explore the most advanced wind generating technologies for improving reliability and decreasing energy costs requires a scaling study. To find the optimum sizes for wind turbines and define size limits, the design process constantly updates and makes improvements as additional data becomes available. As a designer, it's important to revisit the total cost of energy (COE) as wind turbines grow more sophisticated and increase in size.

The impact of increased size and configuration begins in the conceptual design phase developing scaling relationships for subsystems, components, and cost elements. These cross a range of sizes from kilowatt-size turbines to megawatt turbines and pushing upwards to futuristic 10-megawatt turbines in recent years.

Calculating weights for cost modeling can be divided into numerous systems and subsystems with further division into separate costing categories. These are five of the most prevalent categories:

- Rotor blade
- Rotor hub
- Low-speed shaft and bearings
- Gearbox
- Rotor mechanical brake

Rotor Blade

Rotor blades encounter extreme conditions and many variable load forces, which will necessitate extensive investigation. This is best done with scale modeling of many load cases and around three sub-models of the blade—spar weight, aerofoil weight, and flange weight—each having its own separate function.

Blade Spar

Located along the leading edge of the rotor blade, the blade spar behaves as the main support structure of the rotor blade. It also provides support to the protective cladding for both itself and the aerofoil fairing. It also supports the aerodynamic load forces of the blade occurring from fluctuating wind inflow and cyclic bending.

Scaling models for calculating blade loads is complex and requires many case studies to identify extremes load forces. Fatigue and stress from the fluctuating wind and bending forces reduces the capacity of the blade. Testing has shown wind gust loading is the most extreme load a blade has to withstand. Represented by theoretical formulation, the weight of blade spar (W_{BS}) model looks like:

$$W_{BS} = 8.5 \times 10^{-2} F_{CL} F_{RC} \rho_a V_d^2 \lambda_m^2 \left[\frac{1+t}{t}\right]\left[\frac{\rho_{SP}}{\sigma_{SP}}\right] B \left[\frac{D}{2}\right]^3$$

TABLE 5-1 CYCLIC LOAD FACTOR F_{CL}

Hub Type	Blade Frequency Type	F_{CL}
Rigid	Rigid	1.00
Teeter	Rigid	0.85
Rigid	Flexible	0.70
Teeter	Flexible	0.60

Data from Harrison, R., Hau, E., & Snel, H. (2000). *Large Wind Turbines Design and Economics.* West Sussex, England: John Wiley & Sons Ltd.

where ρ_{SP} and σ_{SP} represent the density and strength of the spar material of construction. The cyclic load factor (F_{CL}) gives a comparison of the fatigue characteristics for different blade and hub configurations **TABLE 5-1**. (A **factor**, in this sense, is a statistical variable whose value is independent of the load and weight formula but relevant to the calculated decision being made.)

Cyclic load amplitudes are diminished when the flexibility of a blade is less rigid and the hub is allowed to teeter, as shown in the comparison of hub type to blade frequency type. A teetering hub and flexible blade reduce fatigue, thereby reducing cyclic load amplitudes.

Also along this line of case loading is the rotor control factor (F_{RC}) for showing the effects of different control strategies on the stress of the rotor blade **TABLE 5-2**. Again, the more flexible (variable) the system, the less stress, whereby fluctuating wind gusts can be absorbed by the rotor acceleration.

Aerofoil Cladding

The aerofoil cladding establishes an aerodynamic surface to the rotor blade allowing the least resistance for the lift-to-drag flow needed for rotation of the turbine rotor assembly. This external protective covering gives the rotor blade its smooth, seamless overlay for the overall spar and fairing surface area of the blade.

Modeling of the aerofoil cladding does not take into consideration load stress due to fatigue and bending loads. It is

NOTE

During the initial design phase, the spar support and fairing aerofoil design are closely interrelated. This is not the case when cost modeling begins; the two are disseminated and taken into consideration separately.

TABLE 5-2 ROTOR CONTROL FACTOR F_{RC}

Control Type	Rotor Speed	F_{RC}
Full-span variable pitch	Fixed	1.00
Stall	Fixed	0.85
Full-span variable pitch	Variable	0.80

Data from Harrison, R., Hau, E., & Snel, H. (2000). *Large Wind Turbines Design and Economics.* West Sussex, England: John Wiley & Sons Ltd.

presumed that these stresses are being handled by the spar. For the purpose of modeling, the only function of the aerofoil cladding is to provide a sufficient surface area for the lift-to-drag phenomena to occur. Therefore, the shape and surface qualities become important, but no structural strength is taken into consideration. Determination of the aerofoil cladding weight is estimated by the formulation:

$$W_{BA} = 30F_A \left[1+t\right]S\frac{\pi D^2}{4}$$

whereby W_{BA} is the total blade area weight of the glass reinforced cladding and the blade thickness-to-chord ratio. The weight factor of the aerofoil (F_A) allows for an adjustment in the weight for different materials of construction as given in **TABLE 5-3** .

Blade Root Flange

Located at the base or hub connection end of the rotor blade, the blade root flange borders the rotor blade as a projection for added strength and for bolt attachment of the blade to the hub **FIGURE 5-2** . It also establishes the rotor-to-hub load path, allowing for the transfer of rotational torque and rotor thrust to the hub.

TABLE 5-3 AEROFOIL WEIGHT FACTOR F_A

Aerofoil Material	F_A
Glass reinforced polyester	1.0
Glass reinforced epoxy	0.6

Data from Harrison, R., Hau, E., & Snel, H. (2000). *Large Wind Turbines Design and Economics.* West Sussex, England: John Wiley & Sons Ltd.

FIGURE 5-2 Large wind turbine blades staged and awaiting installation on a rotor hub.
© iStockphoto.com/hudiemm

TABLE 5-4 ROOT FLANGE FACTOR F_{RF}

Root Flange Factor	F_{RF}
Full-span pitch control (conventional)	1.0
Full-span pitch control (advanced blades)	0.5
Fixed hub, rigid blades, stall control	0.14
Teeter hub or flexible blades	0.1

Data from Harrison, R., Hau, E., & Snel, H. (2000). *Large Wind Turbines Design and Economics.* West Sussex, England: John Wiley & Sons Ltd.

The weight calculation for a set of blade flanges is determined by:

$$W_{BF} = 2.1 F_{RF} \left[\frac{\rho_F}{\sigma_F} \right] T_R D^{0.7} B$$

where (W_{BF}) is the weight of the blade flange with ρ_F and σ_F respectively representing the density and strength of the flange materials. Again, the root flange factor (F_{FR}) is given for different control configurations. A full-span pitch control with fixed hub, rigid blades formulates to be the heaviest flange weight calculation while a stall control configuration with the same fixed hub, rigid blades will be the least as shown in **TABLE 5-4**.

Two other important comparisons of blade weight and their respective dynamics are tip-speed ratio (TSR) and specific strength of materials **FIGURE 5-3**.

Depending on the TSR of a wind turbine, the blade component weights will require an increase or decrease in their weight. The relationship of these changes can best be seen in the following comparison chart of each subcomponent to the total weight of a blade **FIGURE 5-4**.

Understandably, when the materials of construction for the spar are changed, the blade weight will change also. Excluding the weight of the aerofoil cladding and the flange, the following figure demonstrates the density comparison of several currently popular spar materials to their respective stress and strength capacity.

Rotor Hub

The rotor hub can be looked upon as the Grand Central Station of load transferral **FIGURE 5-5**. Load forces from the rotor blades to the low-speed shaft (main drive train) must pass through the hub. It is a main structural load path for the tower turbine with various loads being experienced; each having influence on structural weight calculations including:

- **Rotational loads**—Inherent in the blades' ability to capture the wind effectively, these are the direct loads given to the rotating of the blades about

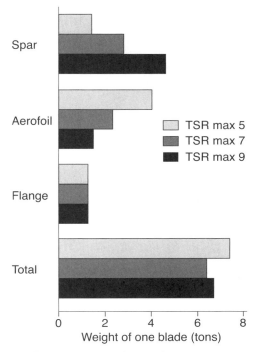

FIGURE 5–3 Effect of TSR on the weight of the blade components.

Data from Harrison, R., Hau, E., & Snel, H. (2000). *Large Wind Turbines Design and Economics.* West Sussex, England: John Wiley & Sons Ltd.

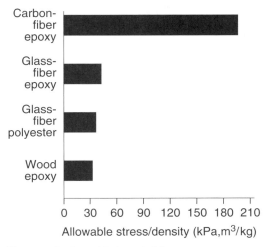

FIGURE 5–4 Specific strength of spar blade materials.

Data from Harrison, R., Hau, E., & Snel, H. (2000). *Large Wind Turbines Design and Economics.* West Sussex, England: John Wiley & Sons Ltd.

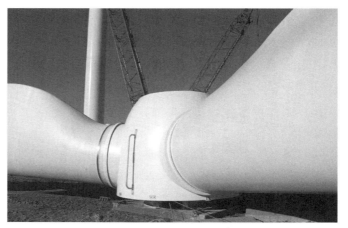

FIGURE 5-5 A hub staged on the ground at a tower site awaiting installation.
© BESTWEB/ShutterStock, Inc.

the horizontal axis of the wind turbine. These rotational loads transfer to the low-speed shaft as high torque, low speed load forces.

- **Flap-wise bending moments**—These account for the aerodynamic force of blade flexibility and stiffness, such as wind shear, and wind fluctuations causing the rotor blades to flex and dampen loads.
- **Edge-wise bending moments**—These account for the gravity loading by the blade itself against the rotational motion of the rotor blade.
- **Gyroscopic loads**—Are experienced when the wind turbine is yawed about its vertical axis.
- **Torsion loads**—Occur at the longitudinal axis of the blade and happen when a change in pitch angle of the blade is made. Torsion loads occur only with pitch-regulated turbines and with aerodynamic braking when installed on regulated turbines using tip brakes.

Each blade portal (arm) of the rotor hub is considered a cantilevered cylinder whereby the rotor blade root bending moment acts upon the hub arm outer rim edge while the inner hub arm is rigidly supported. The maximum blade root bending moment will be assumed and is the product of the dynamic pressure on the rotor blade. With these constraints the weight of the hub structure weight (W_{HS}) formulates as follows:

$$W_{HS} = 42 F_{HG} F_{HL} F_{HC} \rho_a V_R^2 SD^3 \left[\frac{\rho_H}{\sigma_H} \right]$$

where ρ_H is density and σ_H gives allowance for stress of the materials-of-construction. The rotor assembly blade diameter (D) plays a big role in the determination of hub weight due to not only the weight of the rotor blades (H_G) themselves but the forces acting upon them as described by the four load-path forces highlighted above.

TABLE 5-5 HUB GEOMETRY FACTOR F_{HG}

Rotor Type	F_{HG}
3 blade	1.00
2 blade	0.75

Data from Harrison, R., Hau, E., & Snel, H. (2000). *Large Wind Turbines Design and Economics.* West Sussex, England: John Wiley & Sons Ltd.

The number of blades on the rotor assembly influences the structural weight indirectly through their solidity. A three-blade turbine will have a lower TSR of 7 and therefore a solidity of 6.5 percent, whereas a two-blade assembly with a TSR of 10 will have a solidity of 2.7 percent. Applied to the above calculations, a two-blade assembly reduces hub structural weight by close to a factor of two. Also, the number of blades will vary the hub casting size and complexity (three-blade portals being more complex than two-blade portals) for a hub geometry factor (F_{HG}) as shown in **TABLE 5-5**.

The different gyroscopic loadings adjust overall hub weight as indicated by the hub load factor (F_{HL}). A three-blade assembly will give a lower gyroscopic yaw loading than a two-blade assembly due to the moment of inertia (the tendency of the blade to continue moving) of a three-blade being independent of the blade positions. The moments of inertia of a two-blade assembly changes with the position of the blades giving a higher load factor. The use of flexible blades helps reduce these loads as shown in **TABLE 5-6**.

Also influencing size and complexity of the hub is the method of rotor blade control. It stands to reason that the greater the control, the more accommodating the hub must be, as shown in **TABLE 5-7**.

TABLE 5-6 HUB LOAD FACTOR F_{HL}

Rotor Type	F_{HL}
3-blade rigid hub	1.00
2-blade teeter hub	0.75
3-blade with rigid hub and flexible blades	0.75

Data from Harrison, R., Hau, E., & Snel, H. (2000). *Large Wind Turbines Design and Economics.* West Sussex, England: John Wiley & Sons Ltd.

TABLE 5-7 HUB CONTROL FACTOR F_{HC}

Control Type	F_{HC}
Full-span pitch control	1.00
Stall control	0.83

Data from Harrison, R., Hau, E., & Snel, H. (2000). *Large Wind Turbines Design and Economics.* West Sussex, England: John Wiley & Sons Ltd.

A secondary function and component of the rotor hub is the partial housing of the blade pitch bearings and pitch mechanism. This sub-component of the hub is for full-span pitch regulated turbines only and follows the theoretical equation:

$$W_{HPM} = 0.11\left[W_B B + 57 B M_B \left[\frac{\rho_S}{\sigma_S} \right] \right]$$

NOTE

The above equation allows for pitch mechanism weights where individual electric servomotors are used for each blade.

Two terms in this equation for weight calculation of the hub pitch mechanism (W_{HPM}) relate specifically to the pitch mechanism and the bearings.

One last item to include in the discussion of the rotor hub weight is its spinner. This cone-shaped housing couples onto the hub upwind of the rotor blades. It has aesthetic value, but is mostly valued for its aerodynamic purpose of encountering and dispersing the wind first, allowing for the aerodynamic inflow wind stream to be more evenly distributed and less disturbed along the full length of turbine towerhead hub and nacelle housing.

Low-Speed Shaft and Bearings

The low-speed shaft and bearings function as a transferring device for several weight and towerhead system load forces **FIGURE 5-6**. The rotor assembly transfers its forces and loads to the turbine gearbox via this shaft. The low-speed shaft also supports the rotor assembly weight, becoming part of the turbine load path for both gravitational weight and thrust loads onto the bedplate of the nacelle. This load path is set up to occur via the shaft bearings and housing. All this transference of loads via the turbine load path reduces the shear forces on the gearbox.

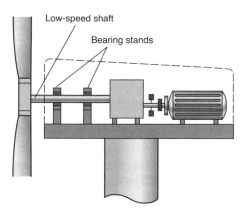

FIGURE 5-6 Illustration of standard turbine configuration depicting location of low-speed shaft and bearing stand.

Adapted from Harrison, R., Hau, E., & Snel, H. (2000). *Large Wind Turbines Design and Economics.* West Sussex, England: John Wiley & Sons Ltd.

Having a low-speed shaft connect the rotor assembly and gearbox also allows for the placement of the gearbox over the tower center line via shaft length correlation. This correlation decreases the towerhead bending moments while also reducing loads on the bed plate and yaw system bearings.

In addition to being a load path, the low-speed shaft can be hollow or solid. As a hollow shaft it becomes a raceway for blade pitch equipment, and hydraulic and electrical control signals from the turbine control panels to the hub.

Weight evaluation modeling of the low-speed shaft is configured as a shaft suspended between two separate standing bearing units.

Shaft dimensions are calculated using a standard method for fatigue of a rotating shaft under applied bending moments known as the *Westinghouse code formula*. The diameter of the shaft is calculated from:

$$d_0 = \sqrt[3]{19.6\left[\left[\frac{Q_{LSS}}{\sigma_y}\right]^2 + \left[\frac{M_{LSS}}{\sigma_e}\right]^2\right]^{1/2}}$$

where σ_y is the stress of the shaft material and σ_e gives the endurance limit. The rated torque (Q_{LSS}) on the shaft is multiplied by a safety factor of 3. The design bending moment for the shaft (M_{LSS}) is multiplied by a safety factor of 1.25 and is calculated as the shaft enters the first main bearing as:

$$M_{LSS} = 0.25 L_{LSS} g W_{ROT}$$

where L_{LSS} is the length of the low-speed shaft. The weight of the rotor assembly (W_{ROT}) is calculated up to the point of its connection flange. Depending upon the bearing-stand location on the shaft will determine length of the moment shaft arm. Shaft weight then is calculated from its dimensions and density.

Gearbox

The gearbox functions to increase the main drive train high-torque and low rpms matching that of the generator shaft speed of low torque and high rpms. There are two types of gearbox configuration that are widely used.

- Parallel—Shaft gears are bulky and heavy, but relatively inexpensive and are the usual choice at lower torque levels.
- Epicyclic (planetary)—Shaft gears are compact and lightweight. In higher nominal torques they are relatively more expensive.

The gearbox weight model uses "p" to designate parallel and "e" for epicyclic. TABLE 5-8 can be used to look up the number and the type of stages (typically two or three) to give the required gear ratio for the design drivers of the particular turbine control type.

TABLE 5-8 GEARBOX TYPE AND OVERALL RATIO

| | Overall Ratios Available in the Model U_O | |
Gearbox Type	Lowest	Highest
p	1.6	6.3
pp	6.3	18.0
ppp	18.0	80.0
ep	8.0	20.0
ee	20.0	40.0
eep	40.0	70.0
eee	70.0	200.0

Data from Harrison, R., Hau, E., & Snel, H. (2000). *Large Wind Turbines Design and Economics.* West Sussex, England: John Wiley & Sons Ltd.

The weight of the gearbox will be calculated from the sum of the weight of each stage with the weight of a single stage formulated by:

$$W_{GSN} = 3.2 Q_S F_S F_W / F_{GD}$$

where Q_S represents the output torque of the stage in question, and service factors F_W and F_{GD} are inserted, respectively, as a weight factor and the effect that the material's properties and surface finishing techniques of the gear teeth have on weight. The gearbox service factors (F_S) are derived from the following listed turbine control configurations **TABLE 5-9**.

Rotor Mechanical Brake

The rotor mechanical brake accounts for less than 1 percent of the total towerhead weight when located on the high-speed shaft after the gearbox. If mounted on the main drive train it could go as high as 2 percent. Its weight is calculated from the sum of the weight from the brake calipers, the brake disk, and the hydraulic pack.

TABLE 5-9 GEARBOX SERVICE FACTOR F_S

Control Type	Rotor Speed	Gearbox Service Factor (F_S)
Full-span variable pitch	fixed	1.75
Stall	fixed	2.00
Full-span variable pitch	variable	1.25

Data from Harrison, R., Hau, E., & Snel, H. (2000). *Large Wind Turbines Design and Economics.* West Sussex, England: John Wiley & Sons Ltd.

Mechanical Equipment

Mechanical equipment entails those items that are difficult to model by the type of formula you have seen in the chapter. They also tend to be small in comparison (5 percent total) and weight modeling is not critical for each. They are typically represented as a whole with each unit developed by choosing an appropriate design driver and then using the weight data available.

- High-speed shaft and coupling—Interfaces the gearbox high-speed shaft to the generator shaft. The design driver for this function is the reduced torque of the rotor blades via the overall gearbox ratio for what is now a low torque, high-speed shaft.
- Lubricating system—In the overall turbine is a centralized oil circulation system feeding the gearbox, main shaft bearings, pitch bearings, and all other items requiring lubrication. The design driver for this oil will be designated by the generator-specified oil.
- Hydraulic system—Provides the high-pressure hydraulic fluid for the rotor mechanical brake and the pitch mechanism if used. The yaw system is also hydraulically driven.
- Air conditioning and fire control—Maintains the environment inside the nacelle within operable temperature for site locations in harsh climates of extreme heat. Automatic fire control equipment will extinguish any fire possibility in the nacelle.

Electrical Generator and Electrical Equipment

The electrical generator and all its peripheral hardware of switch gear, power factor correction capacitor banks, transformers, and cabinets are taken into consideration for modeling and weight calculations. It is calculated that 20 percent of the equipment weight lies in the nacelle at the towerhead with the transformers and cabinets on the ground at the base of the tower.

Generator power cables feed through the inside of the tower and twist as the rotor and nacelle yaw for best wind capturing direction. The power cable length, generator voltage, and current are the design drivers for the power cable. The weight calculation of the cable is derived from tables compiled from manufacturers' data.

The towerhead generator weight (W_{GN}) calculations are dependent upon whether it is an induction or synchronous generator. An induction generator rating (P_g) greater than 275 kW (typically for large commercial wind turbines) is formulated as:

$$W_{GN} = F_{GI} P_g + F_{G2}$$

and synchronous generators regardless of rating the weight are formulated as:

$$W_{GN} = 14.86 P_g^{0.75}$$

TABLE 5-10 INDUCTION GENERATOR CALIBRATION COEFFICIENTS		
Generator Speed (rpm)	F_{G1}	F_{G2}
1,000	4.5	41
1,500	3.13	418

Data from Harrison, R., Hau, E., & Snel, H. (2000). *Large Wind Turbines Design and Economics.* West Sussex, England: John Wiley & Sons Ltd.

where F_{G1} and F_{G2} are calibration coefficients dependent upon generator speed as given in **TABLE 5-10**.

Nacelle and Yaw System

Three main components are taken into consideration in the nacelle and yaw system model.

- Nacelle bedplate—Designed by the placement of turbine processes superimposed upon the bedplate frame. The weight of the frame being the sum of steel required to support the rotor torque (W_{BPQ}), the rotor thrust (W_{BPTHR}), the rotor weight (W_{BORWT}), and the necessary bedplate area (W_{BPAREA}).
- Nacelle cladding—A lightweight housing of glass reinforced plastic (GRP) made in panel construction. It fully encloses the turbine components and provides protection to the tower maintenance crew from the elements. Stiffeners and frame structure are included in the weight calculations of the nacelle housing.
- Yaw system—Weight is calculated as a function of the moment due to the sum of all the weight of the other components in the towerhead. The yaw system itself consists of a slewing ring type gear and bearing assembly with a drive motor and may also have a mechanical brake.

Steel Towers

The steel tower of a wind system design is the most critical support structure in the wind development system. The tower is analyzed as a tubular structure with three main dimensions all related to weight comparisons while meeting strength, stability, and resonant frequency design values.

- Hub height (H_h)—A somewhat fixed dimension needing to be greater ($H_h > D$) than the rotor assembly diameter, understandably, so the rotor blades do not hit the ground when they rotate. This is a simplified explanation and the designer should also incorporate the influence of other important parameters such as energy yield, tower fatigue loads, tower mass, and tower resonant frequency. Tower bending resonant frequency is of utmost importance and can be classified according to its value designated as tower frequency (f_t) whereby three classifications are put into play as:

- Stiff tower: $f_t > BP$ (blade passing frequency, wind shear, yaw error)
- Soft tower: $1P < f_t < BP$
- Soft-soft tower: $f_t < 1P$ (rotational frequency of the rotor)

These classifications are not universally accepted, and depending on the turbine being a variable speed or constant speed, they are debatable.

- **Tower radius (R_t)**—Relates to the tower bending frequency whereby area (I) is given by:

$$I = \pi R_t^3 \delta$$

This gives a cross section of the tower area but you'll also need to formulate the towerhead mass, the rotor swept area, solidity of the rotor blades, the natural frequency of tower height, tower foot (base) bending moment, and wall thickness of the tower. The torque forces on the base of the tower are massive, and considerable thought goes into the design of tower height to tower base radius.

- **Wall thickness (δ)**—The final criteria for tower stability and must be sufficient to withstand buckling of the tower wall. A minimum ratio of wall thickness to tower radius is calculated by:

$$\delta \geq 2R_t / 175$$

For an optimal tower design both strength and buckling must be satisfied simultaneously with a resultant calculation for tower radius and wall thickness such as:

$$R_{to} = \left[\frac{175 M_b}{2\, \pi \sigma_{adm}} \right]^{1/3}$$

with respect to weight and cost. Further and more precise formulation and calculation will go into the design decision of a tower height, radius, and structure final design.

FINDING JOBS IN WIND TECHNOLOGY

All prominent wind industry manufacturers, construction management companies, general contractors, suppliers, and owner/developers will list job opportunities on their websites. If you subscribe to free industry newsletters, you will get to know who these industry players are and even establish where ongoing or start-up projects are happening. There is also nothing wrong with contacting these companies and inquiring about career opportunities.

(Continues)

FINDING JOBS IN WIND TECHNOLOGY (Continued)

There are many Internet sites specializing in renewable and wind energy careers and job opportunities. Most sites will also have a free newsletter or means of obtaining notification regarding upcoming job opportunities. A Web search will bring up many, but several good sources are:

- *www.greenjobs.com*
- *www.renewableenergyjobs.com*
- *www.windenergyjobs.com*

Joining organizations active in the wind industry is of great benefit in keeping you abreast of the latest projects and company involvements. These organizations also offer ongoing programs for educational forums and articles about cutting-edge technologies and design processes. Several such organizations are:

- The American Wind Energy Association (AWEA) *www.careersinwind.com*
- The National Renewable Energy Laboratory (NREL) *www.nrel.gov*
- The American Council on Renewable Energy (ACORE) *www.acore.org/careers*

NREL has many ongoing educational projects and has written materials for study. A very helpful and insightful publication available on the Internet titled "Careers in Renewable Energy" gives many examples of the types of career opportunities available. The article link is *www.nrel.gov/docs/fy01osti/28369.pdf*.

CHAPTER SUMMARY

Validation of a wind turbine design involves complex formulations and calculations over a wide range of variables. Weight loads are the driving design criteria for needing to derive comparisons and predictions to yield load analysis quickly and cheaply. Virtual experimentation offers not only a practical method of research and testing, but also for investigative research to be done before expensive prototype production begins.

KEY CONCEPTS AND TERMS

Factor
Scaling

CHAPTER ASSESSMENT: DETERMINING WIND TURBINE WEIGHT AND COSTS

1. In 1999, the US Department of Energy (DoE) began its Wind Partnerships for Advanced Component Technologies (WindPACT) project.
   - ❑ **A.** True
   - ❑ **B.** False

2. The physical weight of the rotor blades directly affects the bending moment of the:
   - ❑ **A.** gearbox.
   - ❑ **B.** high-speed shaft.
   - ❑ **C.** main drive shaft.
   - ❑ **D.** nacelle bedplate.

3. What bridges the gap between theoretical formulation and the reality of a physically operating wind tower?
   - ❑ **A.** Design drivers
   - ❑ **B.** Design parameters
   - ❑ **C.** Service factors
   - ❑ **D.** Modeling methodologies and tools

4. Which of the following are considered to be modeling tools? (Select two.)
   - ❑ **A.** Virtual experimentation
   - ❑ **B.** Performance predictions
   - ❑ **C.** Scaling models
   - ❑ **D.** Component interfacing

5. Scaling models refers to observable measurements across a wide range of wind turbine operating conditions.
   - ❑ **A.** True
   - ❑ **B.** False

6. The rotor blade weight directly reflects upon what other structural components of the wind turbine? (Select three.)
   - ❑ **A.** Pitch mechanism
   - ❑ **B.** Low-speed shaft
   - ❑ **C.** Generator
   - ❑ **D.** Nacelle bedplate

7. The inverse and byproduct of rotor torque is _____

8. Thrust loading from the rotor assembly affects which downstream components?
   - ❑ **A.** Main drive shaft bearings
   - ❑ **B.** Nacelle bedplate
   - ❑ **C.** Tower
   - ❑ **D.** Tower foundation
   - ❑ **E.** All of the above

9. The rotor blade _____ is the main support structure of a rotor blade and support for the protective cladding of the aerofoil fairing.

10. What dynamic phenomenon does the aerofoil cladding of a rotor blade establish?
    - ❑ **A.** Cyclic load amplitudes
    - ❑ **B.** Rotation of the rotor assembly
    - ❑ **C.** Lift-to-drag
    - ❑ **D.** Fairing surface

11. What load forces influences the size and complexity of the rotor hub?
    - ❑ **A.** Rotational
    - ❑ **B.** Gyroscopic
    - ❑ **C.** Torsion
    - ❑ **D.** Flap and edge-wise bending moments
    - ❑ **E.** All of the above

12. If the load speed main drive shaft is of a hollow style, this allows it to behave as a raceway for what other systems? (Select three.)
    - ❑ **A.** Hydraulic control signals
    - ❑ **B.** Blade pitch control equipment
    - ❑ **C.** Rotor tip speed electronics
    - ❑ **D.** Electrical control signals

13. The cone shaped housing at the tip of the hub is called a(n) _____.

14. The weight calculations of the towerhead generator depend upon whether it is an induction or synchronous generator.
- ❑ **A.** True
- ❑ **B.** False

15. What three main components are taken into consideration for weight modeling of the nacelle and yaw system? (Select three.)
- ❑ **A.** Bedplate
- ❑ **B.** Generator
- ❑ **C.** Nacelle cladding
- ❑ **D.** Yaw system

Weight and Cost of Different Turbine Concepts

WEIGHT AND COST REVOLVE around basic concepts of a fairly standard turbine design. The new generation of wind turbine on the market today requires design drivers in the megawatt range. These advanced designs can compel you to take a second look at more sophisticated configurations and higher capacity formulations. This chapter will examine the comparative advantages and disadvantages of several different wind turbine concepts concerning the rotor blade, main drive train, nacelle, and the tower with reference to weight reduction.

Advanced design drivers provide the opportunity to propel wind turbine efficiency into a higher array of production capabilities. These advanced designs are pushing wind turbine construction from the single megawatt range of power production to upwards of 10 megawatts. Newer, more innovative construction materials and manufacturing methods are coming into play along with control methods and towerhead component redesign for major weight reduction and cost rewards.

In order to evaluate these advanced design drivers without crossing into theoretical designs, this chapter will focus on the standard of the current commercial wind generation industry, i.e., a horizontal axis, three-blade rotor assembly with a 400-foot diameter (200-foot blade).

Chapter Topics

This chapter covers the following topics and concepts:

- Rotor blade materials and stiffness
- Control methods
- The drive train and the nacelle
- Blade configurations
- Direct-drive generators
- Rotor diameter

Chapter Goals

When you complete this chapter, you will be able to:

- Identify advanced design drivers for different towerhead components
- Understand how weight reduction is achieved to obtain cost rewards
- Compare various weight and cost analyses for power output efficiency
- Discuss the weight and cost issues of a vertical axis wind tower
- Discuss wind–diesel production in developing countries' quest for renewable energy
- Calculate the towerhead weight of vertical axis wind turbines

Rotor Blade Materials and Stiffness

It's important to look at the rotor assembly for any design improvements, as this is the workhorse of the wind turbine. Comparison and constraints at the rotor assembly affect all other towerhead component designs the most, and therefore become the first and foremost place to make advances in wind technology. Rotor blades are the powerhouse of the wind turbine, with the materials of blade construction and considerations of blade stiffness being the most important for advanced concepts.

This chapter will first look at comparisons of a three-blade rotor configuration **FIGURE 6-1** and blades made from glass-reinforced polyester. Blade length is assumed to be approximately 200 feet and it is steel flanged at its root-hub connection. A conventional blade of this nature weighs on the average of 6.5 tons each, with the spar contributing 45 percent of the weight, the fairing 35 percent, and the root flange 20 percent.

The current trend to build modern-day wind towers at megawatt capacity can be achieved only by putting into service larger and longer turbine blades. Advanced construction materials offer the most opportunity to improve wind turbine efficiency by giving a weight advantage **FIGURE 6-2**. This advantage allows for the

FIGURE 6-1 Classic example of a commercial horizontal axis three-blade wind turbine.
© Hemera/Thinkstock

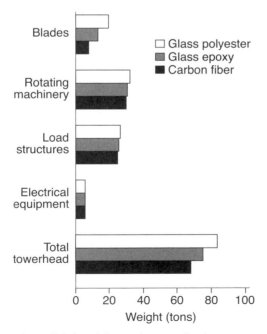

FIGURE 6-2 Comparison of blade weight to other towerhead components with three different materials of construction.
Data from Harrison, R., Hau, E., & Snel, H. (2000). *Large Wind Turbines Design and Economics.* West Sussex, England: John Wiley & Sons Ltd.

larger, longer, stronger blades required for megawatt production. These higher-level blades, in turn, cover a greater swept area, effectively increasing the blade tip-speed ratio to further exploit the capture of raw wind energy for a more profitable bottom line.

The tip-speed ratio (ratio of blade tip speed to the speed of the wind) for a high efficiency three-blade turbine works out to a ratio factor of between six and seven. Keeping blade weight down and under control then becomes an important goal, since blade-mass determination is calculated as the cube of the blade radius. Weight loading due to gravity is a major constraint factor for rotor assemblies using the larger megawatt blades. Reduced blade weight also helps to reduce the blade base flange weight.

Epoxy-based composite blades generate the greatest interest for designers in the wind industry. This is due to the combination of environmental, production, and cost advantages beyond the conventional polyester used in early designs or the more recent resin-type construction systems.

Epoxy-based composites allow for shorter cure cycles, increased durability, and improved surface finish. The two most current and sought-after epoxies are as follows:

- Glass–fiber epoxy—Used for the blade spar and aerofoil. This material of construction also allows for an improved flange design. Reduction in total blade weight can now be realized at around 4.4 tons with the spar contributing 55 percent of total weight, the aerofoil outer skin at 30 percent and the blade root flange at 15 percent.

  With blade weight reduction, additional weight-reducing methodologies carried over into the towerhead itself can be realized. Pitch mechanism, low-speed shaft and bearings, mechanical brake, and nacelle bedplate are all reevaluated for weight-reducing advantages.

  Less weight in the blade means fewer bending moments on the drive train and less load force. Self-weight of the rotor blades becomes more meaningful the larger the turbine diameter, with another 2.1 tons of weight reduction for an overall reduction of approximately two tons or 30 percent as compared with 100 percent fiberglass.

- Carbon–fiber epoxy—Recently has been identified not only as a cost-effective means for reducing blade weight but increasing blade stiffness as well. Carbon fiber is estimated to result in a 38 percent reduction in total blade mass and a 14 percent decrease in cost when compared with a 100 percent fiberglass design. An added benefit is the reduction of the thickness of the fiberglass laminate sections and an advanced flange or "stud" connection at the blade root.

  The blade weight reduction is similar to that of glass-fiber epoxy with a single-blade weight reduced to 2.6 tons accompanied again with reductions

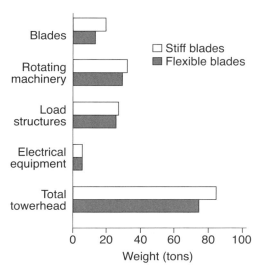

FIGURE 6-3 Comparison of blade stiffness on towerhead weight.

Data from Harrison, R., Hau, E., & Snel, H. (2000). *Large Wind Turbines Design and Economics.* West Sussex, England: John Wiley & Sons Ltd.

in the same towerhead component weights. In addition to the weight reduction, carbon-fiber epoxy has a more environmentally friendly reputation and is therefore becoming a leading contender for use by wind industry manufacturers regardless of its higher cost.

- **Flexible blades**—Allow for load reduction forces on the blade spars, the blade root flange, and the hub, resulting in respective weight reductions **FIGURE 6-3**. Weight reduction of the blade also allows for similar reduced weight at the hub structure, hub mechanism, main drive shaft, and accompanying towerhead components (nacelle, bedplate, etc.) for a total of 4.5 tons. This is in addition to the reduction realized for glass-fiber epoxy and carbon-fiber materials of construction already realized.

Lowering the towerhead weight clearly offers substantial benefits in costs. However, these are offset by the cost of highly expensive composite materials and their extraordinary cost of production during the manufacturing of the blades. The offset does not exceed weight benefits and therefore a great deal of emphasis is being placed on prototype materials and methods of production.

One more area of developmental intrigue is the performance benefits to using composite materials in blade manufacturing. The newer style materials have proven to accelerate the rotor assembly more readily due to their lower rotational inertia. As the wind speed picks up or fluctuates with wind gusts the tip-speed ratio remains more constant, allowing the turbine to operate closer to its tip-speed ratio more of the time. This improves energy capture and thus leads to a more steady output. This is something the wind farm works toward achieving for annual energy output ratings to the utility companies.

Control Methods

Control of the turbine rotor assembly is done for numerous reasons, but mainly to coordinate its speed and torque to stay within the tolerances of the rest of the turbine components. The main drive shaft and the generator both have tolerances within which they must operate. This allows them to function at their maximum efficiency and to minimize fatigue or overloading issues. The power grid the wind farm intends to supply wind generated electricity to, also has set requirements that must be considered.

The rotor assembly you set your comparisons around will be a fixed speed, full-span, variable pitch control method. This control method will be compared to that of a stall control and a variable speed control turbine **FIGURE 6-4**.

- Stall control—Will have weight reduction in some components but an increase in others, such as the blade spars, which are designed to be heavier. However, there is an offset, because the pitch mechanism is not needed.

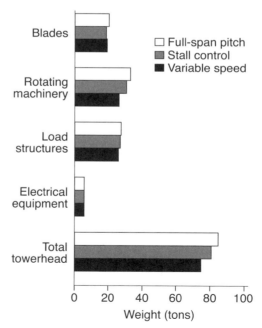

FIGURE 6-4 Comparison of three styles for the rotor assembly control and their effect on towerhead weight.

Data from Harrison, R., Hau, E., & Snel, H. (2000). *Large Wind Turbines Design and Economics*. West Sussex, England: John Wiley & Sons Ltd.

Also, the blade root flange and hub structures are reduced in weight. An increased service factor requires a heavier gearbox and a blade-tip brake is added. There is an overall weight reduction, giving the stall-control method the advantage, in terms of weight, over the fixed speed, full-span, variable pitch control method.

The total rotor assembly weight reduction with this control method is about nine tons, while the gearbox weight increase is about four tons for an overall reduction in weight to five tons total. Further advances in stall control of the rotor assembly show a lower dynamic load spectrum than our comparison rotor assembly using pitch control. This points toward further weight savings than indicated.

▪ **Variable speed**—Rotor assembly control will reduce rated torque and reduce the service factors on the low speed shaft and the gearbox thereby allowing a reduction in the weights of these components. In turn, this allows reduction of the gearbox support structure and the nacelle bedplate. Most important, variable speed allows the reduction in weight of the rotor blades themselves. The turbine generator can also be resized to a lower profile.

Variable speed control can add up to substantial weight savings, but these savings are offset by the expensive electrical equipment and controls needed to operate such a system. This makes the variable speed control almost as costly as the fixed speed rotor assembly. However, the much more effective energy capture of the variable speed control system brings the final unit costs of electricity production in line to make this a viable system for consideration.

Drive Train and Nacelle

Compactness of the drive train and the nacelle is a current trend of study in the wind industry. Integrating the drive train, its gearbox, and the turbine generator will result in a lighter, more compact system, with fewer gears and bearings. This design style is about two tons lighter in weight than a conventional drive train. Also, the reduced weight and space requirements in turn reduce the nacelle bedplate.

The towerhead drive train and nacelle you will use for comparison is the modular arrangement. This style of arrangement consists of the high-torque, low-speed drive shaft emerging from the hub and supported by two separate bearing mounts. A modular style drive train will dictate the length of a conventional chassis, namely the nacelle bedplate, and will be compared to a more compact drive train and accompanying bedplate **FIGURE 6-5**.

Weight reductions will be realized with these advance design approaches by not only the mechanical equipment involved but also for the bedplate and nacelle.

Classical bedplate mounted drive train

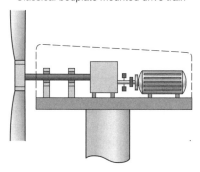

Compact drive train concepts

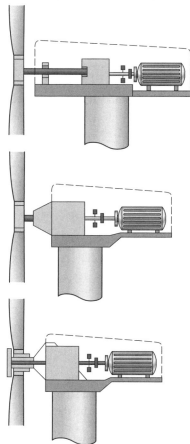

FIGURE 6-5 Examples of various compact drive-train configurations.

Adapted from Harrison, R., Hau, E., & Snel, H. (2000). *Large Wind Turbines Design and Economics.* West Sussex, England: John Wiley & Sons Ltd.

Cost savings are less dramatic, as the bedplate is already at a relatively low cost. A 20 percent reduction in the bedplate weight reflects a cost savings of only 6 percent.

To summarize the design features discussed so far, while only modest savings may be realized for individual design changes, a combination of these design concepts produces substantial reduction in weight with accompanying cost rewards. Comparing a conventional wind turbine of basic design with that of an advanced variable controlled turbine realizes some very dramatic weight reduction. Our advanced turbine for comparison comprises glass-fiber epoxy materials of construction, flexible blades, and a short integrated drive train. This comparison, as demonstrated in **FIGURE 6-6**, reflects a total weight reduction of 44 percent and a reduction in total cost of 20 percent.

Blade Configurations

There are relative advantages to having a different number of blades on a rotor assembly but that is not the only way to achieve increased performance and reduction of overall tower weight. Tip-speed ratio (λ_d) and its effect on the coefficient of performance (C_p) for the wind tower require crucial consideration.

Each addition of a blade to the rotor assembly decreases the tip-speed ratio. Where tip-speed ratio is assumed to be at maximum coefficient of performance; the coefficient of performance decreases with each increment of blade ratio factor. This relationship can be seen best with the following table comparing the tip-speed

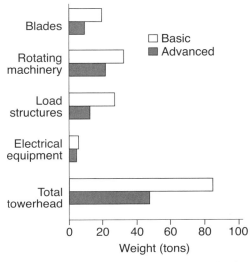

FIGURE 6-6 Comparison of basic turbine to advanced weight reduction measures and the effect on the total wind farm development.

Data from Harrison, R., Hau, E., & Snel, H. (2000). *Large Wind Turbines Design and Economics*. West Sussex, England: John Wiley & Sons Ltd.

TABLE 6-1 COMPARISON OF TIP-SPEED RATIO AND COEFFICIENTS OF PERFORMANCE		
Machine	λ_d	C_P
Three blade	7	0.45
Two blade	10	0.43
One blade	15	0.4

Data from Harrison, R., Hau, E., & Snel, H. (2000). *Large Wind Turbines Design and Economics.* West Sussex, England: John Wiley & Sons Ltd.

ratio of the three most standard blade numbers to the coefficient of performance **TABLE 6-1**.

This relationship is significant as an increase in tip-speed ratio gives a reduction in the levels of the rotor torque. Less rotor assembly torque affects all downstream turbine components having to deal with the rotation loads of the rotor assembly. The main drive shaft, gearbox, bearing mounts, and bedplate are directly affected. The generator, nacelle, and tower are indirectly affected due to the tower load path for a total overall weight reduction **FIGURE 6-7**. Also the hub structure is reduced due to fewer blade attachments and mechanisms for pitch control.

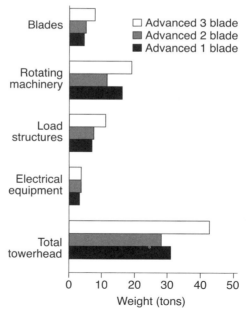

FIGURE 6-7 Comparison of advanced blade configurations in terms of weight reduction and the effect on the total wind turbine development (based on the three-blade turbine having a fixed hub and the two-blade turbine requiring a teetering hub).

Data from Harrison, R., Hau, E., & Snel, H. (2000). *Large Wind Turbines Design and Economics.* West Sussex, England: John Wiley & Sons Ltd.

Offsetting these weight advantages is the need for a teetering hub on a two-blade rotor assembly and a counterweight on a one-blade assembly. The concept of using a two-blade and one-blade rotor assembly receives much research and development (R&D) attention due to the advantage of weight reduction. However, the downside of lower coefficient of performance, higher noise emissions, and lack of aesthetic value has overshadowed their widespread commercial use.

Increasing tip speed also increases the need for blade spar reinforcement or increased weight. Hence the one-blade weight advantage is negated by the need for a heavier blade spar and its counterweight making the weight value negligible.

The two-blade basic fixed turbine and three-blade variable advanced concept show towerhead weights to be comparable **FIGURE 6-8** . Now, you need to take the costs of each turbine into consideration. Costs of the basic two-blade turbine are in line with the three-blade advanced. Once the two-blade turbine gains in design maturity and production, it may well have the cost advantage.

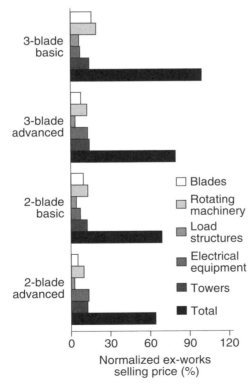

FIGURE 6-8 Comparison of a three-blade and a two-blade rotor and the effect on the selling price of the ex-works (the total external components used by the wind turbine).

Data from Harrison, R., Hau, E., & Snel, H. (2000). *Large Wind Turbines Design and Economics*. West Sussex, England: John Wiley & Sons Ltd.

Direct-Drive Generators

The main advantage of direct-drive generators is they do not contain a gearbox and consequently the modular-style main drive shaft is shortened. The generator of a direct-drive approach is mounted directly behind and supported by the main drive shaft, and positioned where the gearbox traditionally sits **FIGURE 6-9**.

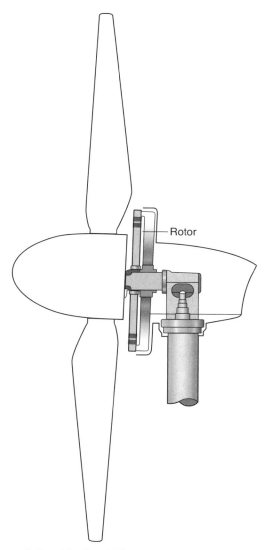

FIGURE 6-9 Enercon E-40 turbine (500 kW) with direct-drive generator, housed in the frontal lobe of the nacelle.

Adapted from Harrison, R., Hau, E., & Snel, H. (2000). *Large Wind Turbines Design and Economics.* West Sussex, England: John Wiley & Sons Ltd.

This approach and style of a generator typically will weigh more than the combined weight of a gearbox and high-speed generator of the same power rating. The total weight will make the nacelle both heavier and more expensive in manufacturing. This is where modeling tools come into play for weight and cost comparisons.

The three comparative configurations to be addressed are:

- Conventional geared drive train with modular drive train and high-speed generator
- Shortened drive train with single-stage gearbox and synchronous generator
- Direct-drive synchronous generator

The model used to assess these three approaches is a horizontal axis wind turbine with a 200-foot diameter rotor assembly, rated at 1.5 MW. The conventional fully geared drive train formulates out to be considerably heftier with a mass 25 percent greater than that of the other two approaches, and more expensive besides. The intermediate approach of a shortened drive train and single stage gearbox showed no appreciable weight or cost savings. The third approach is the most current trend of investigation but has several challenges.

The motivation for using a direct-drive generator is increased efficiency and avoidance of gearbox issues of high maintenance and reliability. The challenges become those of achieving a generator that is lighter and more cost effective than that of the conventional approach, all the while pushing turbine output limits. The driving force of this technology is the higher-speed, larger-sized turbines climbing upward to the 10 MW range and beyond.

Most gearbox issues are associated with the high-speed gear stages of megawatt turbines. So the question you need to ask yourself is at what point is the weight-to-efficiency limit reached and when do the comparisons hit their maximum. New hybrid turbines are being researched every day for just such maximums.

Most current direct-drive technology is based on use of permanent magnet synchronous generators. This approach requires additional electronic control and monitoring, adding both weight and space issues. As these systems are being developed, further study is being conducted on peripheral components to bring weight down. The driving force to this style of generator system is to achieve the higher outputs (10 MW and higher) being demanded of the wind industry.

Rotor Diameter

The key dimension for optimizing size and subsequent power output of a wind tower is the length of the rotor diameter. This dimension constitutes the core challenge of future wind turbine design. It sets the baseline that determines all other turbine components.

The current trend is to increase the diameter of the rotor assembly, because this makes it possible to capture more wind, but at the same time it also increases the weight and cost of the towerhead. The question asked of designers when modeling and making design comparisons is how to allow this incremental increase of the rotor diameter while keeping the weight and cost under control. Where can weight and cost be trimmed for each incremental increase in rotor diameter?

The following is a weight and cost model, along with the comparison of basic and advanced blade design for both a two- and a three-blade turbine. This comparison is set against the towerhead weight (W_T) with the rotor diameter expressed as an exponential function:

$$W_T = kD^n$$

whereby the coefficient (k) and the diameter exponent (n) both vary somewhat over the rotor diameter size range (100 to 200 feet) but the average value of the weight exponent stays the same at 2.7. This has been verified and remains consistent as derived from statistical data from actual operating wind turbines.

The percentage of weight contribution by the various towerhead components over the rotor blade diameter comparison range (100 to 200 feet) remains fairly consistent. The tower weight may increase slightly in the lower range but levels off and stays consistent when compared against the other towerhead components.

When variations in the specific costs of the turbine are charted against the rotor diameter, it is expressly identified that the two-blade turbine is the most cost effective **FIGURE 6-10**. The three-blade turbine is consistently more expensive over the whole rotor diameter size range. It should be noted that the weight range of a three-blade turbine actually overlaps that of a two-blade turbine. It is when a

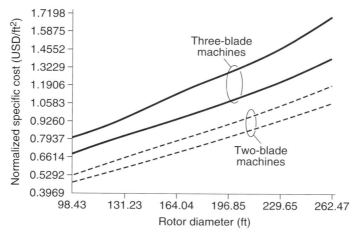

FIGURE 6-10 Charted increase of specific cost of two- and three-blade turbine against model rotor diameter range.

Data from Harrison, R., Hau, E., & Snel, H. (2000). *Large Wind Turbines Design and Economics*. West Sussex, England: John Wiley & Sons Ltd.

cost factor is added that the two-blade turbine is consistently less costly. This is because of the weight gains of towerhead structure, such as the nacelle and bed-plate, are comparatively inexpensive.

Of course increasing rotor diameter and subsequently the hub height of the tower is hopefully expected to produce an increase in wind energy production. How that computes out is a major consideration, as illustrated in the following chart. It is the reverse of what might be expected.

Where tip speed and rotor speed increase with the lengthening of rotor diameter, the towerhead weight and costs rise slightly more steadily than the expected increase in energy capture. This is the designer's and manufacturer's dilemma—how to get weight and costs down in comparison to incremental energy capture.

Also there is a tip speed for which the maximum power coefficient of the wind tower can be obtained. Tip-speed principles dictate the rotor assembly rpm allowable for maximum energy output. The tip speed is therefore kept constant over the course of blade lengthening because of these underlying principles. This computes to the low-speed, high-torque load forces on the main drive shaft.

The two design drivers most important when considering diameter length to weight and cost are as follows:

NOTE

Tip speed and tip-speed ratio are not the same. Whereas tip speed is designed to be held constant resulting in higher torque loads as it increases, tip-speed ratio is the comparison of the rotor blade tip's speed to the speed of the wind.

- **Rotor torque**—The dominant design driver. As the rotor blade is lengthened and the rotor blade tip speed is held in check, the rotor torque increases substantially, and thus weight factors of the towerhead to withstand this torque increases **FIGURE 6-11**.

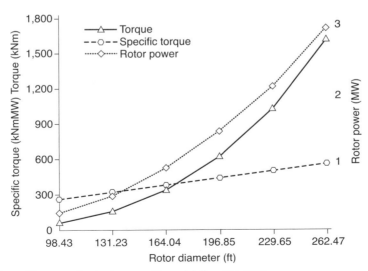

FIGURE 6-11 Charted increase of rotor torque with model diameter range.

Data from Harrison, R., Hau, E., & Snel, H. (2000). *Large Wind Turbines Design and Economics*. West Sussex, England: John Wiley & Sons Ltd.

■ Rotor self-weight—This gives rise to the three load types (gravitational, aerodynamic, extraction), again giving rise to towerhead load forces and the accompanying structural weight to withstand these loads.

These prediction models thus far do not take into account the price trends for the **ex-works**, the external components used by the wind turbine but not part of it, such as the tower structure, electrical equipment on the ground, and overhead power lines. Nor do the models consider that manufacturing methods change as wind turbines become larger. These trends can be charted and predictions made, but as actual prototypes are built and value engineering applied, such statistics will change.

Calculating Towerhead Weight of Vertical Axis Wind Turbines

Vertical axis wind turbines **FIGURE 6-12** have not been as popular in the commercial arena as horizontal axis turbines. They are, however, being considered more seriously for smaller power production plants, such as on commercial and residential buildings. The vertical axis allows for greater leeway in blade design and mounting locales making them more convenient for individual and private use. For these reasons, R&D for vertical axis turbines is on the rise with many and varied blade designs, often very ingenious.

This chapter will focus on the conventional H-configuration vertical axis **FIGURE 6-13**, with information currently available on the weight and costs but

FIGURE 6-12 Vertical axis H-rotor concept wind turbine.
© eldeiv/ShutterStock, Inc.

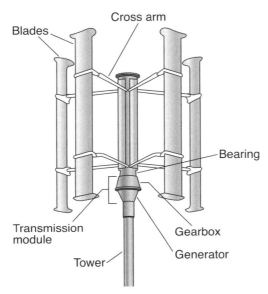

FIGURE 6-13 Representative drawing of vertical axis H-rotor concept wind turbine.
Adapted from Harrison, R., Hau, E., & Snel, H. (2000). *Large Wind Turbines Design and Economics.* West Sussex, England: John Wiley & Sons Ltd.

not fully representative of all vertical type turbines. Our comparison model divides the H-turbine into seven subcomponents:

- **Blades**—On a stall-regulated vertical axis turbine they are complex. For modeling purposes, it is assumed that maximum loads occur at the stalling point and that the blade spar is a tapered rectangular box of glass-reinforced polyester.
- **Cross arm**—This design driver is the bending movements produced by its self-weight and aerodynamic forces. Modeling criteria assume the cross arm to be a tapered steel box spar surrounded by an aerofoil of glass-reinforced polyester.
- **Brakes**—Mounted on the low-speed side of the gearbox. Heavy calipers and disks are used due to exceptionally large braking forces.
- **Main bearing**—This design driver is the vertical and horizontal forces on the bearing and the number of revolutions executed during the turbine's lifetime.
- **Low-speed coupling**—Connects the cross-arm hub to the main shaft of the gearbox. This is a simple torque tube with thickness calculated from the allowable stress in the tube.
- **Gearbox**—A three-stage epicyclic with a service factor of three, allowing for the high cyclic loading of a vertical axis turbine.

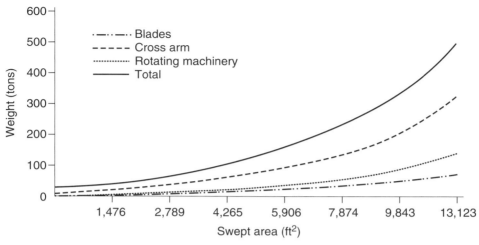

FIGURE 6-14 Charted comparison of towerhead weight to swept area for an H-configuration vertical axis wind turbine.
Data from Harrison, R., Hau, E., & Snel, H. (2000). *Large Wind Turbines Design and Economics*. West Sussex, England: John Wiley & Sons Ltd.

- Power transmission module—Contains and supports the power transmission equipment. It is a steel structure, housing the brakes and bearings; its weight is calculated from rotor blade dimensions and loads.

    The majority of the towerhead weight comes from the heavy cross arms and increases steadily with increase in swept area as noted in **FIGURE 6-14**. For future advances in the H-rotor turbine, the cross arm is the key component for addressing weight reduction and the high-drag restriction preventing higher tip-speed ratios.

WIND–DIESEL SYSTEMS IN DEVELOPING COUNTRIES

Developing countries dependent on expensive imported diesel fuel for generation of electricity are getting a boost into the wind industry. In Central Europe, some wind farms have been around long enough that they're ready for a new generation of turbine technology. This process of upgrading the technology of a wind farm to boost its output is known as **repowering**. It makes for more efficient wind farms but also often results in a lot of second-hand turbines on the market at relatively low prices.

The used towers are being offered at attractive enough prices to warrant their reuse elsewhere. Exporting these towers to developing countries is under analysis for both technical and economic feasibility with one wind facility already in operation. The recycling of used wind turbines will supplement the increasingly expensive imported diesel fuels for remotely populated areas.

WIND–DIESEL SYSTEMS IN DEVELOPING COUNTRIES (Continued)

One such site chosen for introduction of wind energy is the remote town of Gao (40,000 inhabitants) in northern Mali. Gao is located on the southern rim of the African Sahel. Gao gets its electricity from a 4-MW diesel-fueled generator system. A wind project of three towers, 300 kW each, was built as a supplemental resource for the diesel fuel, which has to be imported and transported over a distance of more than 1,500 miles.

In contrast to wind farm projects that are developed in close proximity to a national power grid, a remote wind-diesel system is much more interactive, keeping its produced power localized. These isolated grids are between 1 kW and 10 MW, with storage facilities such as batteries or thermal dump loads installed. The localized power grid for the Gao project covers an area of six by nine miles, with a peak load of approximately 1.8 MW from 120-foot towers and moderate wind speeds. The existing diesel power station is 4 MW, with the wind farm supplying about 15 percent of the annual electrical demand.

Due to its small size and high infrastructure costs, the Gao wind farm investment costs were almost double that for a commercial development with national grid connections. However, it is still feasible that such a technological and economical concept will be transferred to many other remote regions of the Sahel as well.

CHAPTER SUMMARY

This chapter delved into the weight and cost of the new megawatt wind turbine. These turbines require considerably more design process, with design drivers capable of more sophisticated configurations and higher capacity formulations. You learned advantages and disadvantages of several established model comparisons of the rotor blade, main drive train, nacelle, and the tower, always with the intent of weight reduction and cost rewards.

Advanced design drivers using modern materials of construction and manufacturing methods are pushing the wind turbine industry into higher and higher megawatt production capabilities. The key factor in this is the rotor diameter. As it becomes larger, the process of balancing load forces and weight becomes more complex. You have taken a second look at the new generation of wind turbine, and it is headed for the megawatt range.

KEY CONCEPTS AND TERMS

Ex-works
Repowering

CHAPTER ASSESSMENT: WEIGHT AND COST OF DIFFERENT TURBINE CONCEPTS

1. Advanced design drivers for more sophisticated turbine configurations and higher capacity formulations have comparative advantages and disadvantages for which wind turbine components?
   ❑ **A.** Rotor blade
   ❑ **B.** Main drive train
   ❑ **C.** Nacelle
   ❑ **D.** Tower
   ❑ **E.** All of the above

2. Currently, the US commercial wind generation industry is fairly standardized around which two design drivers? (Select two.)
   ❑ **A.** Horizontal axis
   ❑ **B.** Three-blade rotor assembly
   ❑ **C.** Solid concrete towers
   ❑ **D.** Direct-drive generators

3. Conventional blades on average weigh 6.5 tons each, with the spar contributing _____ percent of the weight, the fairing _____ percent, and the root flange _____ percent.

4. Larger, longer, stronger rotor blades for megawatt production cover a greater swept area, effectively increasing the blade tip-speed ratio to further exploit the capturing of raw wind energy.
   - ❑ **A.** True
   - ❑ **B.** False

5. Which composite blade materials of construction generates the greatest interest for designers in the wind industry due to the combination of environmental, production, and cost advantages? (Select two.)
   - ❑ **A.** Glass-reinforced polyester
   - ❑ **B.** Resin-fiber epoxy
   - ❑ **C.** Carbon-fiber epoxy
   - ❑ **D.** Glass-fiber epoxy

6. Flexible blades can give additional weight reduction by reducing load forces on which of the following component(s)?
   - ❑ **A.** Spar
   - ❑ **B.** Root flange
   - ❑ **C.** Hub
   - ❑ **D.** Main drive shaft
   - ❑ **E.** All of the above

7. An increase in tip-speed ratio gives a reduction in the level of rotor noise.
   - ❑ **A.** True
   - ❑ **B.** False

8. Control of the turbine rotor assembly is done for numerous reasons but mainly to coordinate its _____ and _____ to stay within the tolerances of the rest of the turbine components.

9. A current trend in the wind industry integrates which of the following for a lighter, more compact towerhead? (Select three.)
   - ❑ **A.** Main drive train
   - ❑ **B.** Gearbox
   - ❑ **C.** Bedplate
   - ❑ **D.** Generator

10. An increase in tip-speed ratio also increases the level of rotor torque.
    - ❑ **A.** True
    - ❑ **B.** False

11. The motivation for using a(n) _____ generator is increased efficiency while avoiding gearbox issues of high maintenance and reliability.

12. What is the key dimension for optimizing size and thus power output of a wind tower and constitutes the core challenge of future wind turbine design?
   ❑ A. Direct-drive generator
   ❑ B. Compact towerhead
   ❑ C. Rotor diameter
   ❑ D. Blade aerodynamics

13. On a vertical axis H-style wind turbine the majority of the towerhead weight comes from the heavy _____ and increases steadily with increase in swept area.

14. Currently, the commercial wind generation industry is fairly standardized around the horizontal axis, three-blade rotor assembly with a 400-foot diameter (200-foot blade).
   ❑ A. True
   ❑ B. False

Wind Turbine Siting, System Design, and Integration

THIS CHAPTER COVERS SITING—the process of deciding where to locate a wind farm. You'll learn to examine it on a national and local level. Also, you'll delve into the wind facility's social and economic impact on its immediate surroundings.

There is more to developing a profitable wind transmission facility (wind farm) than building towers. Selection of a transmission location site, the facility's transmission systems, and the incorporation of that system into a national power grid with all its policies and regulations must be contemplated very seriously before a developer begins to build a wind farm. This process can, and most often will, last many years before actual construction begins. Not only are meteorological test sites monitored for several years for determination of available and viable wind resource, but developers also meet with local, state, and federal groups to discuss environmental impact during this time as well.

The infrastructure required to get wind energy from the turbine to the national power grid requires advance planning and conformance with national, state, and local regulations and policies. The system design requires explanation and planning, as does the integration into the national power grid with its own set of regulatory policies.

This chapter goes beyond the wind turbine itself and looks at how and why a particular wind transmission facility site is chosen, how the system supporting that tower siting is designed for optimum power production, and lastly, how the overall system design will be integrated into an already established national power grid.

Chapter Topics

This chapter covers the following topics and concepts:

- Local, state, and federal policies along with regulatory measures for wind turbine siting
- The operation of a wind transmission facility, including its integration into a national power grid

Chapter Goals

When you complete this chapter, you will be able to:

- Discuss how the different policies and regulatory statutes at the local, state, and national levels affect the siting of a wind transmission facility
- Discuss the elements of federal policy that protect the environment, wildlife, and vegetation near a wind transmission facility
- Understand the effect that each level of policy and regulatory agency has on the siting of a wind turbine facility

Overview

The **siting** of a commercial wind farm requires consideration of several factors critical to a successful project. The following criteria are taken into account over and above the actual wind farm towers and support infrastructure facilities. They are:

- Land that has attractive wind resources, as established via meteorological test towers with a minimum of two years' data, and is available for a wind farm—either as open land or land whose other uses are compatible with a wind farm, e.g., agriculture
- Convenient accessibility, both during construction and over the course of the facility's operations.
- Close proximity to a national power grid and a suitable electrical interconnecting point
- Feasibility permitting of both landowner and community support, along with participating utilities' assurance that additional transmission spur lines have a high degree of recoverable investment; the whole project has to make obvious sense to all parties involved

* Favorable electricity market with a rate of return on investment (ROI) in a predictable manner
* State and federal regulatory approval
* Aviation compatibility

Utility-scale wind energy facilities impact the local communities in which they are built in many ways. Many communities establish ordinances addressing safety, land use, and other issues before wind farms are developed in order to regulate the impacts of wind farms.

Wind Turbine Siting

A wind transmission facility can be anywhere from one or two towers to upwards of several hundred wind turbines all connected to a utility grid at one location. No matter the number of towers per wind site, once a location is selected, it will immediately become a local, state, and even national concern.

Much study and wind data collection has gone into the process of determining the wind velocities throughout the United States and its shorelines for the explicit purpose of siting wind transmission facilities. This data has been collated into a national map, showing the location of land-based and offshore wind resources **FIGURE 7-1**. A great deal of environmental study goes into the most advantageous placement of a wind farm.

The highest, most consistent yield of wind resources is offshore, mainly on the New England coast. The permitting and developing of these winds is currently under consideration and moving forward.

A strong and consistent wind energy resource is not the only factor in siting a wind transmission facility. Several factors come into play before a developer begins siting a wind farm.

From the US wind resource map it is clear the majority of land-based wind resources are positioned in the Midwest and Great Plains region. However, tapping into these high-yield resources may not be the most economically advantageous option.

Many wind farms are being built in remote areas of Arizona. Arizona does not have the highest wind yields, but it does have access to enormous power-grid transmission lines due to hydroelectric dams such as the Hoover Dam on the Arizona/Nevada border. National power grids are located there, delivering electrical power to major populations. Another location is the Columbia River Gorge, which runs between the bordering states of Oregon and Washington. This has been a major wind development area for many years. The area has both the wind

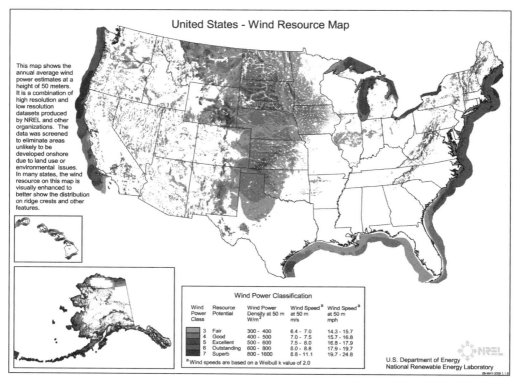

United States - Wind Resource Map

This map shows the annual average wind power estimates at a height of 50 meters. It is a combination of high resolution and low resolution datasets produced by NREL and other organizations. The data was screened to eliminate areas unlikely to be developed onshore due to land use or environmental issues. In many states, the wind resource on this map is visually enhanced to better show the distribution on ridge crests and other features.

Wind Power Classification

Wind Power Class	Resource Potential	Wind Power Density at 50 m W/m²	Wind Speed [a] at 50 m m/s	Wind Speed [a] at 50 m mph
3	Fair	300 - 400	6.4 - 7.0	14.3 - 15.7
4	Good	400 - 500	7.0 - 7.5	15.7 - 16.8
5	Excellent	500 - 600	7.5 - 8.0	16.8 - 17.9
6	Outstanding	600 - 800	8.0 - 8.8	17.9 - 19.7
7	Superb	800 - 1600	8.8 - 11.1	19.7 - 24.8

[a] Wind speeds are based on a Weibull k value of 2.0

U.S. Department of Energy
National Renewable Energy Laboratory

FIGURE 7-1 Wind resource map showing land-based and offshore wind data used for siting wind transmission facilities in the United States.
Courtesy of NREL

potential and electric grids from numerous large hydroelectric dams, thereby making the area highly sought after for wind developers.

The best meteorological wind data, as demonstrated in the wind resource map does not, however, always indicate where a wind farm makes the most economic sense. Quite often, local and state incentives invite a developer to choose sites where communities are open to wind energy development. Federal grant monies, tax incentives, and state funding all play a role in the siting of a wind facility. Also, developers need to consider where a site can be located in proximity to a national grid.

Once a developer collates the data and chooses the best location for a wind farm, every level of government and environmental organization will have a hand in its siting, either through policy guidelines that establish the scope of the project, or regulatory sanctions requiring authorization and a permit process.

Wind siting **policies** at the federal, state, and local level set policy guidelines and regulatory requirements addressing such issues as potential obstruction of airspace, radar interference, and the protection of wildlife resources, along with impact on nearby communities.

Federal Policies

Federal involvement in wind turbine siting involves numerous policies diversified among the management of several different executive departments. Each different agency will have its own viewpoint on the various concerns for the land and welfare of the people.

Bureau of Land Management (BLM) Wind Development Policy

The BLM establishes the criteria by which national public lands are employed in conjunction with the harvesting of available wind resources FIGURE 7-2. The BLM estimates there are approximately 20.6 million acres of public lands with wind development potential. Rights of way given for energy production sites are estimated to have an installed capacity of 327 MW or roughly 2 percent of total currently installed US capacity. Developers have utilized the Midwest and Great Plains states the most because of their abundant wind resource. More important, the BLM gives rights of way for transmission lines crossing these public lands with no wind facilities in the immediate area or public site.

The BLM completed its first analysis for wind energy development in June 2005, calling it the Programmatic Environmental Impact Statement (PEIS). This

FIGURE 7-2 Wind farm development is compatible with other land uses such as agriculture.
© BHamms/ShutterStock, Inc.

statement allows for the authorization of wind energy projects on BLM land. It is used as an aid in analyzing the impact for specific applications requesting use of public lands for wind energy use.

In 2006, a set of best management practices (BMPs) was issued outlining a wind energy policy for the BLM. Its purpose was to help reduce any potential impact on birds, wildlife habitat, and other natural resources. The 2006 policy addresses rental rates, visual resource guidance, requirements for plans of developments, and exclusionary areas.

As an increasing number of applications for rights of way are submitted, the BLM continues to try to improve the management of energy resources on federal public lands.

For assurance of quality of life for both the resource and those in need of the land's resources, a wind energy Plan of Development (PoD) must be submitted at the end of a three-year site testing and monitoring authorization program. This provides the basic information the National Environmental Policy Act (NEPA) process and analysis for a wind project to proceed.

The following is a PoD outline of the minimum requirements taken into consideration for every wind facility development:

- Project description—Information on the type of facility and generation capacity, with a statement of the purpose and need for the project, is required. This is a lengthy and detailed document covering such items as legal land location, total acreage, number and size of wind turbines, substations, water usage, erosion control, vegetation treatment, fire protection, site security, interconnection to power grid, and more.
- Facility construction details—Includes the wind turbine design, geotechnical studies, access and transportation, work force numbers, site preparation, wind turbine assembly, site stabilization, etc.
- Related facilities and systems report requirements—Entails the transmission system, meteorological towers, and other communications systems—such as microwave, fiber optics, hard wire, or wireless—needed during construction, and operation.
- Operation and maintenance (O&M) needs—Refers to both the maintenance activities and work force including labor, equipment, and ground transportation.
- Environmental considerations—Includes general site characteristics and potential environmental issues of sensitive species and their habitats, special land use, cultural and historic sites, recreation conflicts, visual management, aviation or military conflicts, and any other possible issue of environmental concern.
- Maps and drawings—These show a footprint of the facility to include reference to the Public Land Survey already established. These maps and

drawings of the facility layout are typically at 30 percent of the finished engineering and civil design drawings to include site grading plan, access and transportation layouts, and visual resource evaluation and simulations.

As you can see, a great deal of regulatory permitting is required to develop and build a wind facility on public lands. Additional supplemental materials for the NEPA analysis is now required. This is a four-tier information submission consisting of:

- Final engineering and civil design for the facilities, including facility survey and design drawing standards, aviation lighting plan, watershed protection, and erosion control with final site grading plans.
- Alternative considerations of site evaluation criteria, advocate alternatives considered but not carried forward, and comparative analysis of site configurations.
- Facility management plan that addresses storm-water pollution, hazardous materials, waste, invasive species, and noxious weeds as well as health and safety (OSHA requirements), environmental inspection, and compliance monitoring plan.
- Facility decommissioning plan for reclamation and site stabilization, temporary reclamation of disturbed areas, removal of towers, and infrastructure plan in case of future disuse of the wind farm.

Department of Defense (DoD) and Department of Homeland Security (DHS) Joint Program Office

Policy letters were issued in 2006 and 2007 in relation to the siting of wind turbines. Both the DoD and DHS stated they were not opposed to the development of wind farms. They went on to explain further that they would work closely with the Federal Aviation Administration (FAA) case by case to determine if any project posed a potential adverse impact on military or other national security operations. It was stated where a project did raise concerns, the DoD and DHS would approach the appropriate regulatory authority in order to prevent any perceived adverse effect.

To accomplish their intent, the DoD formed the Wind Farm Action Team. This team monitors the establishment of any wind farm development within radar line of site for the National Air Defense and Homeland Security Radars. In 2010, this intent was put to the test when a proposed Oregon state wind farm fell within this criterion and the DoD stepped in to evaluate. With the cooperation of local, state, and federal officials, the wind siting was approved and allowed to move forward within the year.

US Fish and Wildlife Service and Wind Energy Development

In 2002, the US Fish and Wildlife Service and Wind Energy Development established the Wind Turbine Siting Working Group. Their agenda was to develop a set of comprehensive national guidelines to help protect wildlife resources and assist in avoiding post-construction environmental upsets.

It is through the National Environmental Policy Act that the Fish and Wildlife Service became involved in the review of potential wind farm development on public lands. This relationship cooperates to address the Migratory Bird Treaty Act, the Bald and Golden Eagle Protection Act, and the Endangered Species Act. Also, if requested, the Fish and Wildlife Service may become involved on private lands because of its technical expertise in wildlife. This is on a voluntary basis only.

The most recent contribution to enhancement of wildlife and natural resources preservation came with the establishment of the Wind Turbine Guidelines Advisory Committee. This committee provides recommendations for developing effective measures to avoid or minimize impacts to both wildlife and their habitats. These recommendations are to be used by the Fish and Wildlife Services to develop voluntary wind energy guidelines.

US Forest Service

The US Forest Service is responsible for managing 193 million acres and has issued over 74,000 special use authorizations covering over 180 different land uses. It works in very close conjunction to the BLM to monitor and give guidance for the wind industry. In 2007, the Forest Service amended its internal agency directive in order to encompass and give specific guidance to wind energy developments on National Forest System (NFS) lands. The directives are intended to ensure consistent analyses of wind energy proposals and the issuing of permits.

The guideline intention is to monitor wind energy sites before, during, and after construction with regard to wildlife, scenery, significant cultural resources, and national security by such means as:

- Minimum area permit—Allowing a site testing and feasibility permit covering no more than five acres for a single meteorological tower (MET)
- Plan of development document—Describing proposed wind facility, how it will be constructed, operated, and decommissioned
- Project area permit—Allowing a site testing and feasibility permit covering more than five acres for construction, operation, and maintenance of multiple METs
- Significant cultural resource document—Describing national historic landmarks and archaeological objects that are important to the public or scientific community on the proposed site
- Site plan—Scaled and two-dimensional graphic representations of all proposed wind towers, buildings, service areas, roads, and site boundaries along with existing site features such as topography, vegetation, landscape, water bodies, and elevations

FIGURE 7-3 Example showing string-style siting of wind towers.
© F1online/Thinkstock

- **Species management concerns**—Listing of threatened and endangered species including wildlife, fish, birds, or rare plants
- **String layout**—A number of wind turbines oriented in lines or grids in close proximity to one another such as along a ridgeline **FIGURE 7-3**. The close proximity of each turbine has an effect on vegetation in the immediate environment. A new concept of clustering towers **FIGURE 7-4**, rather than siting them in long lines, is becoming more the norm, thus reducing the effects of downstream wind current patterns.

FIGURE 7-4 Example of cluster-style siting of wind towers.
© Pedro Salaverria/ShutterStock, Inc.

This is just a partial listing of the guideline application and consideration process required. There are over 15 sections in total with numerous subsections to be addressed when a developer is looking to locate upon public lands.

National Telecommunications and Information Administration (NTIA)

The NTIA plays the role of resolving technical telecommunication issues for both the federal government and private sector. With respect to the siting of wind turbines, the NTIA addresses interference in radio, microwave, radar, and other frequencies critical to our lines of communication.

When the NTIA receives a proposal from a wind energy developer, it passes it along for comment to the Interdepartment Radio Advisory Committee (IRAC). When an IRAC representative raises a concern, direct discussion will be used to resolve the issues. All this is done on an informal basis with no formal application, recordkeeping, or permitting process.

Federal Regulations

You just learned about federal policies intended to direct the proposed wind development in a direction most suitable for the land and its protection. Now you'll discover the federal **regulations** that developers must adhere to before constructing a wind facility.

The FAA has the most interest and is the major concern for wind development on a federal level. Any structure higher than 200 feet must be evaluated by the FAA. Applications must be submitted 8 to12 months before construction begins, with heavy civil penalties for each day not in compliance. Protecting the safety and improving the efficiency of our national airspace is an FAA priority. Identifying objectives and strategies to satisfy competing demands for airspace while enhancing the safety of its constituents requires the FAA to monitor and permit.

The three areas of greatest concern in the shared utilization of airspace are: marking and lighting, airport encroachment/zoning, and radar interference. Towerhead lights are the most prevalent means of satisfying FAA concerns. Radar interference is also a major consideration when siting a wind development, and if the development is near an airport, the permitting process escalates accordingly. Federal regulations are limited mostly to FAA requirements deferring to state and local policies for more extensive application requirements, with each state setting its own directives individually, as you'll read below.

State and Local Policies

State and local policies are the biggest concern of wind developers. Without the cooperation of local authorities, landowners, and generally positive public opinion, the project most likely will not continue forward. **FIGURE 7-5** shows installed wind capacity by state.

Because wind energy is an electrical generation system, it is regulated by the electrical codes and designed to meet all National Electrical Code (NEC) requirements.

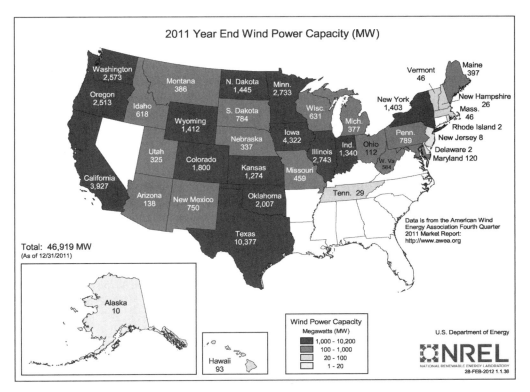

FIGURE 7–5 Installed wind capacity by state.
Courtesy of EERE

There may also be some local and state codes, but generally they are easily met. It is the local and state zoning restrictions that offer the most challenges. These codes are driven by safety concerns but also by the desire of the immediate community to protect their view and that of the horizon landscape view.

Each state and local community has its own set of rules and regulations. These can be addressed by checking with the American Wind Energy Association (AWEA) website. The following are a few applicable standards, rules, and codes relating to wind energy to look for:

- **Efficiency standards**—Placed on equipment with minimum standards of operation. The federal government also provides minimums on certain equipment. The federal efficiency standards prevail over the state standards, even if the states are stricter.
- **Building energy codes**—Apply to commercial wind facilities also. While many local government entities have written their own energy codes, many use existing codes and supplement with state-specific amendments. In some

cases local building energy codes also require commercial facilities to meet green building standards as well.

- Green power purchasing—Government entities such as schools, municipalities, and government buildings can purchase renewable energy credits (RECs).
- Interconnection standards—The technical and procedural process by which a wind facility connects to an electricity power grid. The power grid level will dictate the authority setting the standards such as the Federal Energy Regulatory Commission (FERC) sets the standards for interconnection to interstate transmission lines. A power grid confined to within a state will be state regulated.
- Renewable portfolio standards (RPS)—Require utilities to transmit renewable energy credits (REC) for a certain percentage of electrical sales (or generating capacity) according to a set schedule.
- Wind and solar access laws—Written to establish a right to construct. These laws give the utilities the directive to buy wind- and solar-generated electricity.
- Wind and solar permitting standards—At the local level many areas have adopted simplified or expedited permitting standards for wind and/or solar projects. A *top-of-the-stack* (fast track) concept is put into place, saving time and money in the permitting process. Some states have capped fees as well.

Overall, state and local regulatory setting authorities are not simply allowed to impose whatever requirements they choose. Under federal law established in 1978 via the Public Utility Regulatory Policies Act (PURPA), no publicly held generating facilities can be prohibited from interconnecting to a public power grid. The state and local interconnection requirements cannot be arbitrary and cannot be designed to protect special interest groups other than for safety and reliability.

Installation and Operation

The installation and operation of a wind development is not determined by tower cost alone. The site facilities and power grid connection encompassing capital costs along with operation and maintenance costs are significant. In the megawatt-size range of wind towers (over 200 foot rotor diameter at 3 MW and above) the specific costs for civil construction of the wind site and the electrical equipment are about the same. The developer takes into consideration eight distinct categories when planning the installation costs of a wind facility over and above the cost of the tower and its components:

- Site preparation—Dependent upon the immediate geography of the wind development siting. Preparation costs are influenced not only by location but the condition of the site and can have a more substantial influence on the total development cost than the cost of the towers. The most pivotal condition is that of access roads for tower erection and later maintenance of the site. Also, developers must consider how far the site is from the main thoroughfare for the entrance access to the site.

- Foundation—Construction varies with the manufacturer and height of tower but most importantly it is dependent upon the soil conditions of the site. Usually the foundation will figure at 5 to 9 percent of the total cost. If ground conditions are weak or shifting, then piling must be used adding additional costs of 30 percent in most cases.

- Erection and commissioning—Typically included in the cost of the tower manufacturing and not broken out separately. If erection and commissioning are itemized, their cost can be about half that of the foundation cost.

- Grid connection—Includes the infrastructure transmission lines from the tower siting to a facility substation and power grid transfer station. This cost of installation is driven by the distances from individual towers to the collection transmission lines and the distance and terrain the collection lines must take to the facility's substation location. Tower intermediate transformers and switching gear also figure into the grid connection costs.

- Remote monitoring—Costs increase with the development and use of more highly sophisticated electronics. Nearly all wind farm developments now employ monitoring and control devices and consider it to be a necessity regardless of the cost.

- Financing—The cost of a wind farm can be crucial and add up to a sizable figure. This would include bank fees (for example credits and interest charges during construction) and permit and application costs that can account for 3 percent of the total cost of construction.

- Planning and engineering—Includes negotiating and supervision. These costs are somewhat independent from the size of the project. Regardless to the project size, these cost are incurred no matter if the project goes forward or not. They represent a considerable amount not given to a percent of the total project cost but upfront investment costs.

- Transport—These costs can be very important as they are not inclusive with the purchase of equipment from the manufacturer. Often, equipment needs to be transported over long distances and difficult terrain. In remote sites these issues can make the siting prohibitive or the use of a larger turbine too costly.

Operating and maintenance costs are those expenditures associated with the annual upkeep of the wind farm facility. Management, insurances, and in some cases even taxes are tabulated as operating costs. The main cost figures are usually divided into two categories.

- Preventive maintenance—A regularly scheduled maintenance plan usually carried out via service contracts with licensed service and technician crews that can either be individual companies or through the manufacturer. Scheduled maintenance is usually done at intervals of:

- Six months—Minor repair and inspection of blades, electrical components, lubrication to bearings, linkage, and replacement of worn parts such as brake pads, oil filters, etc.
- Twelve months—Detailed inspection of blades, generators, and gearboxes
- Five years—Major overhaul of blades, generator, and gearboxes

- Repairs—Constitute those times a wind turbine must be shut down. The most frequent of these is the scheduled stoppage for maintenance. Machine-related problems are the largest unscheduled stoppage. Unexpected stoppages such as lightning strikes and grid problems due to storm damage are rare. Stoppages due to generator, gearbox, drive train, rotor, or yaw system faults can cause major shutdowns, even though they occur rarely if maintenance is completed on a regular basis. Electrical and mechanical problems, on the other hand, account for close to half of failure shutdowns with the control system and sensors representing a large portion of the failures FIGURE 7-6.

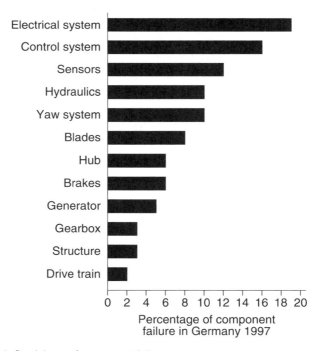

FIGURE 7-6 Breakdown of component failures.

Data from Harrison, R., Hau, E., & Snel, H. (2000). *Large Wind Turbines Design and Economics.* West Sussex, England: John Wiley & Sons Ltd.

Other operating costs that are ongoing and critical to the continued success and profitability of a wind farm are:

- Insurance—This covers the facility operators for accident liability and for major mechanical breakdowns to include loss of income.
- Land costs—These are incurred for annual lease or rent payments to the landowner if on private land. The same is true if the wind site is on federal, state, or BLM regulated land.
- Project management—This is critical if construction changes hands and new management personnel take over. This constitutes about 1 percent of the projected ex-works selling price.

TWENTY PERCENT WIND ENERGY BY 2030: INCREASING WIND ENERGY'S CONTRIBUTION TO US ELECTRICITY SUPPLY

An increase in the goal for US wind energy contribution set at 20 percent by 2030 represents an ambitious growth that would require a buildup of new wind facilities at 3 GW per year from 2010, to more than 16 GW per year from 2018 on **FIGURE 7-7**. Demand for electricity is projected to grow by 39 percent for a total of 5.8 billion MW by 2030. The 20 percent

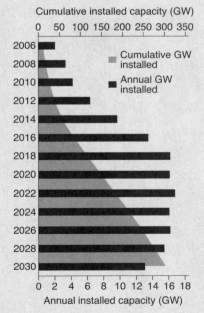

FIGURE 7-7 Annual and cumulative wind installations projected to 2030.
Data from NREL

(Continues)

TWENTY PERCENT WIND ENERGY BY 2030: INCREASING WIND ENERGY'S CONTRIBUTION TO US ELECTRICITY SUPPLY (Continued)

wind scenario not only will have to increase production output to reach the goal but must also increase accordingly to match the overall growth of electrical usage.

A dramatic growth is occurring currently from a very small base (see Figure 7-7) and the fact that wind energy technology has established itself firmly confirms a rapidly growing industry, albeit with varying challenges, impacts, and uncertainty. A reliable mix of wind resources, estimated land needs, required utility and transmission infrastructure, and manufacturing requirements all figure into the pace of growth necessary to meet the 2030 challenge.

The greatest challenge to meeting the 2030 goal is the variability and uncertainty of energy production from wind farm technology. With extensive experience and data collected, utility engineers in some parts of the United States are now helping to reduce these concerns. The expansion of the US transmission grid is underway not only to provide access to the best wind resource regions but also to help relieve current congestion. This will include new transmission lines for the transfer of wind-generated electricity from high resource areas to high demand centers.

Increased wind development has the potential to reduce the need for new coal and combined cycle natural gas **FIGURE 7-8** . However, it would need additional combustion turbine

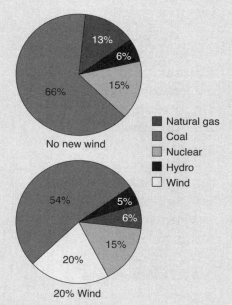

No new wind

20% Wind

Legend:
- Natural gas
- Coal
- Nuclear
- Hydro
- Wind

FIGURE 7-8 Comparison of different fuels.

Data from EERE

TWENTY PERCENT WIND ENERGY BY 2030: INCREASING WIND ENERGY'S CONTRIBUTION TO US ELECTRICITY SUPPLY (Continued)

natural gas capacity to ensure electric grid system reliability, when necessary. Wind, unlike many traditional power sources, cannot be accessed on demand (i.e., turbines produce electricity only when the wind blows). This is called a *non-dispatchable* power source and requires the backup of other systems.

The 20 percent wind scenario is purported to be feasible even with the challenges given. It would require continued evolution of our current transmission power grid along with planning and system operation expansion—all the while maintaining reliable service and reasonable electricity rates throughout the decade and into 2030.

CHAPTER SUMMARY

In this chapter you have gone beyond the wind turbine and its various components to look more closely at the preplanning and regulatory requirements at the federal, state, and local levels concerning the placement of a wind facility. You also learned that selecting the site of a wind transmission facility must be contemplated very seriously before you begin to build a wind farm. This process can last many years.

Going beyond the wind turbine itself, you looked at how and why the proposed site for a wind transmission facility is chosen. This chapter also discussed the various agencies and their policies as they relate to the welfare of wildlife habitat, natural resources, historical and archeological sites, and the overall preservation of the landscape. Lastly, this chapter looked at how a wind site must comply with regulations concerning national security, radar interference, and general atmospheric environment.

KEY CONCEPTS AND TERMS

Policies
Regulations
Siting

CHAPTER ASSESSMENT: WIND TURBINE SITING, SYSTEM DESIGN, AND INTEGRATION GOALS

1. The process of deciding where to locate a wind transmission facility is called _____.

2. A developer must seriously contemplate which of the following before building a profitable wind farm facility?
   - ❏ **A.** Selection of a transmission location site
   - ❏ **B.** A facilities transmission systems
   - ❏ **C.** Incorporating the transmission system into the national power grid
   - ❏ **D.** Complying with local, state, and federal regulatory agencies
   - ❏ **E.** All of the above

3. Which of the following criteria is taken into account over and above the actual wind farm towers and support infrastructure facilities? (Select three.)
   - ❏ **A.** Attractive wind resources
   - ❏ **B.** Convenient site accessibility both during construction and continued operations
   - ❏ **C.** Close proximity to a national power grid and a suitable electrical interconnecting point
   - ❏ **D.** Visually attractive siting

4. The majority of land-based wind resources lie in the _____ and _____ regions.

5. The Bureau of Land Management monitors and sets policy for the protection of which of the following? (Select three.)
   - ❑ **A.** Wildlife habitat
   - ❑ **B.** Natural resources
   - ❑ **C.** Airspace
   - ❑ **D.** Natural vegetation

6. The highest, most consistent yield of wind resources is offshore, mainly on the New England coast.
   - ❑ **A.** True
   - ❑ **B.** False

7. High yield wind resources are not the only reason for the siting of a wind facility. The developer also looks at which of the following? (Select two.)
   - ❑ **A.** Populated areas
   - ❑ **B.** Close proximity to high transmission power grid lines
   - ❑ **C.** Areas with good tax incentives
   - ❑ **D.** Least infringement upon natural resources

8. The four main departmental agencies setting federal policy are the:
   - ❑ **A.** BLM.
   - ❑ **B.** FAA.
   - ❑ **C.** US Forest Service.
   - ❑ **D.** DoD and DHS joint program.
   - ❑ **E.** US Fish and Wildlife Service.

9. Environmental considerations by the BLM program of development (PoD) address potential environmental issues for which of the following?
   - ❑ **A.** Sensitive species and their habitats
   - ❑ **B.** Cultural and historic sites
   - ❑ **C.** Recreation conflicts
   - ❑ **D.** Visual management
   - ❑ **E.** Aviation or military conflicts
   - ❑ **F.** All of the above

10. A number of wind turbines oriented in close proximity of each other in a long line, such as along a ridgeline, is called _____ versus the technique of _____ towers.

11. Because wind energy is an electrical generation system, it is regulated by the electrical codes and designed to meet all State Electrical Code (SEC) requirements.
    - ❑ **A.** True
    - ❑ **B.** False

12. The federal government has both policy guidelines and regulations that must be applied for and met before construction of a wind facility can take place.
- ❏ **A.** True
- ❏ **B.** False

13. State electrical codes take precedence over federal codes.
- ❏ **A.** True
- ❏ **B.** False

14. Any structure higher than _____ feet is required to be evaluated by the FAA.

15. The installation and operation of a wind development is determined by tower cost alone.
- ❏ **A.** True
- ❏ **B.** False

Environmental Impact of Wind Systems

ANYWHERE FROM 10 to 25 percent of wind projects are delayed due to concern over adverse environmental impacts. First and foremost is land use or, more important, the large expanse of land that a wind development requires. Then there is the visual impact of a wind development, which can be formidable, depending on a person's individual viewpoint. Noise emissions produced by the tower's turbine blades and electromagnetic interference are also considered. A proposed wind transmission facility may be delayed or not built at all due to numerous environmental concerns.

Avian and bat mortality rates within and around a wind development are critically important, not only for the conservation of wildlife but for public acceptance. Environmental stewardship programs work toward minimizing the risks of degradation of wildlife ecosystems. These programs promote research and data collection to find solutions that benefit wildlife habitat, thereby enhancing solutions to reduce the impact of a wind system on its immediate environment.

Reducing the risk from wind power developments for loss of habitat and key species is important to agencies, designers, manufacturers, and developers. Removing the barriers to wind power development and increasing the acceptance of wind power technologies require addressing numerous siting and environmental issues. Well thought out and properly sited wind projects can provide an overall net sustainable environment.

Chapter Topics

This chapter covers the following topics and concepts:

- Wind development land-use issues and their immediate impact on surrounding areas
- Avian and bat interaction with wind towers and turbines
- The visual impact of wind towers
- Noise issues
- Electromagnetic interference of wind turbines
- Wind tower safety considerations

Chapter Goals

When you complete this chapter, you will be able to:

- Discuss a wind development's impact on land use, wildlife, and fragile ecosystems within close proximity of the wind site
- Relate the three-tiered classification of wind industry impact to wildlife stewardship
- Understand the impact a wind facility has on commercial, military, and security air traffic control
- Relate safety issues to the structure and surrounding grounds of a wind tower

Land Use Impacts

The overall magnitude of land required to develop a wind energy facility can be sizable in comparison to the actual **footprint** of the tower foundations, roads, and site infrastructure. The footprint adds up to only 2 to 5 percent (28 to 83 acres per megawatt) of the total project area. This is an area that can encompass anywhere from a few hundred to upwards of thousands of acres. However small in proportion, land-use impact is still undeniably an issue to be addressed. The crucial influences of geography, tower layout, blade turbulence, and archeological impact must be accounted for and observed.

Geography can be either open land, where clear access to the wind is available, or forested regions. Open land allows for the use of established farmland without requiring any land use changes. The Midwestern prairies and farming communities with their high wind yields are good examples of open land as an

FIGURE 8-1 Herd of cattle grazing near a wind tower.
© iStockphoto/Thinkstock

ideal choice for siting a wind farm, as are the open BLM rangelands. It is common to see animals grazing in close proximity of a wind tower **FIGURE 8-1**. Building a wind farm in a forested region involves much greater impact on the land, since trees must be cleared to build tower foundations, access roads, power lines, and support buildings.

Tower layout has changed in recent years as larger towers have come on the market. Larger towers make it possible to have fewer towers per site with wider spacing between each tower for the same power output. Towers aligned in rows are referred to as *strings* and are the most common array pattern to date **FIGURE 8-2**.

FIGURE 8-2 Example of a wind farm development where towers are situated in a string or array alignment.
© Photoroller/ShutterStock, Inc.

FIGURE 8–3 Example of cluster-style wind farm tower layout.
© Richard Thornton/ShutterStock, Inc.

Recently designers have started using a *cluster* arrangement of towers **FIGURE 8–3**. This allows for greater possibility in both site selection and best land use placement. It also means the developer can make use of the best wind resources on a wind site.

Blade turbulence is created by the vortices off the trailing edge of each turbine blade, which has an overall effect downstream of the tower. Current data suggest that a commercial wind farm of several to hundreds of wind towers creates a temperature change and a drying effect downwind of its footprint. The moisture content of the soil can be disturbed, which may cause farmers to have to irrigate more. As the larger swept area of the MW-size towers becomes more prevalent, this issue is being given more thorough investigation.

Archaeological and historical resources are often found on wind sites. Assessment for these findings is part of the siting evaluation process. When these resources are found, they are cordoned off and left undisturbed. While the construction and operation of a wind facility can impact its immediate natural surroundings, the degree and type of influence has proven to be site-specific. Each proposed development and wind siting will have its own set of criteria and data in relation to the land use impact within its inherent boundary and, most critically, its footprint.

Larger mammals, rodents, ground-covering carnivores, and grazing animals do not appear to be affected by the wind towers, and in fact they are often found grazing around the wind tower base. In high-temperature climates, herds of wild animals can be seen grazing in the shadow cast by the wind tower.

Avian and Bat Interaction

Avian fatalities, or deaths of birds, as well as deaths of bats, happen when wind facilities are developed in the migratory flight path or feeding airspace aloft or their established habitats closer to the ground. (**Avian** means "related to birds.")

Over the years, both developers and government agencies have thought wind turbine blades were related to avian and bat fatalities but recent research has produced some interesting findings.

Fatalities from contact with a wind tower or its turbine blades are not the major cause of avian and bat deaths. Various research data suggest that other causes lead to much higher incidence of mortality to these wildlife species, as indicated in FIGURE 8-4. That said, diminished numbers in any amount of a given species are taken seriously by wind industry proponents, and major concern is warranted.

Bat fatalities are of special concern, as many species are already showing signs of decline at a time when wind developments are on the rise. Bat populations are particularly sensitive to mortality rate changes, tending to recover very slowly following any incident of population destruction. Wind turbine–related deaths are of major concern to the wind industry.

As the wind industry expands and the number of wind development sites continues to increase, the number of fatalities from wind turbines is expected to increase relative to other structures of incidence. Federal laws addressing wildlife protection in reference to wind development impact are:

- The Migratory Bird Treaty Act (16 U.S.C. 703-712), as amended
- The Bald and Golden Eagle Protection Act (16 U.S.C. 668-668d), as amended
- The Endangered Species Act (16 U.S.C. 1531-15440)

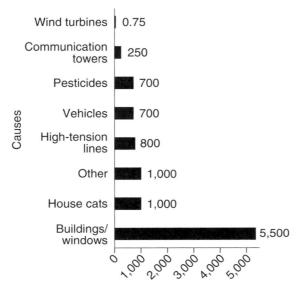

FIGURE 8-4 Comparison of causes of avian fatality in the United States.
Data from EERE

Population decline of many bird and bat species due to habitat loss and fragmentation, other invasive species, and human interference warrants close consideration and efforts to help these species by every means possible. Increasing awareness of each species' patterns of behavior helps tremendously in the effort to diminish their mortality rates. Many species, especially bats, which have low reproductive rates, are shown to recover from population declines very slowly and require special consideration.

Since the mid-1990s, extensive research and data collection has taken place at numerous wind sites representing different types of terrain and ground cover. While tower blades do not contribute as substantially to avian fatalities as was previously proposed, collaborative partnership programs continue to look for better and more efficient means of protecting all wildlife. This is done through such peer-based organizations as:

- The National Wind Coordinating Collaborative
- The Grassland and Shrub-Steppe Species Collaborative
- The Bats and Wind Energy Cooperative

These and other organizations use research and data collection for a better understanding of how to avoid, minimize, and make less severe the impact of wind developments. By study of the encroachment on the **flora and fauna** of our untamed natural environment, they elevate and set new standards. In May 2009 nearly $2 million in environmental research grants was awarded to study reducing the risks to key species and habitats.

As wind-generated power gains acceptability, the commercial wind turbine is increasing in size to obtain greater efficiency and power production. As wind turbine size increases, the blades will become wider and longer—posing a greater potential risk and threat to birds and bats.

The most used methodology for estimating wildlife fatalities is simply counting the animals. This method has been used to account for fatalities by numerous causes such as disease, environmental accidents (oil spills, radiation leaks, etc.) or human structures (power lines, buildings in general, cars). Wind turbine field studies use the same method for determination and estimation of fatality percentages around tower sites. However, with this method also comes imperfect detection of carcass counts. This gives rise to inaccurate estimates.

Fatality percentages must be adjusted before numbers are finalized. Among factors that skew the numbers of carcasses counted:

- Removal of carcasses by scavengers before researchers arrive
- Researchers' inability to recover all carcasses due to variety of ground conditions
- Search protocol of sampling intervals, methods of collection, weather and turbine allowable inference, seasonal elements, etc.

As shown in **FIGURE 8-5**, documented mortality rates at most facilities vary and are site specific. There are no averages or typical scenarios predicting how a

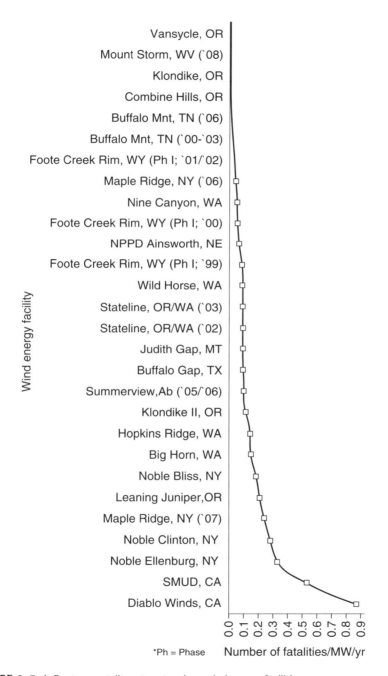

FIGURE 8–5 A. Raptor mortality rates at various wind energy facilities.

Data from the National Wind Coordinating Collaborative

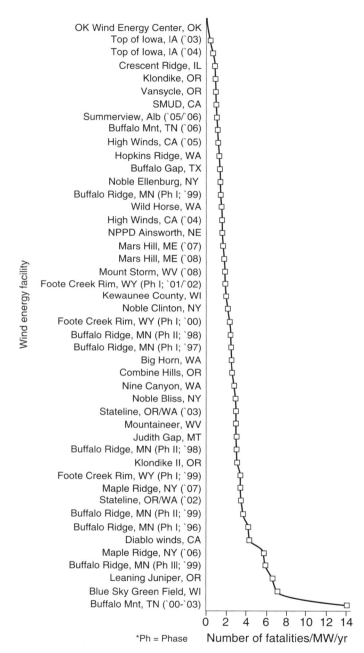

FIGURE 8–5 B. All bird mortality rates at various wind energy facilities.

Data from the National Wind Coordinating Collaborative

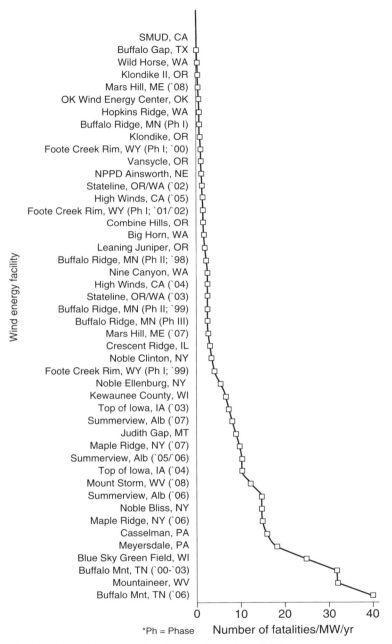

FIGURE 8–5 C. Bat mortality rates at various wind energy facilities.

Data from the National Wind Coordinating Collaborative

wind farm development will impact its surrounding environment and wildlife. Each site will have its own set of variability and environmental dilemmas to comprehend, study, and deal with. There are, however, a few notable characteristics that can be gleaned from fatality charting of many wind sites.

Mortality rates of **raptors**, or birds of prey, are relatively low, with their death toll less than one incident per year for each megawatt of power output. This compares with fatalities of all types of birds averaging in the single digits for each megawatt of power output, and bat deaths numbered in the teens per megawatt of output. One exception to this observation is in California, where older and smaller turbines do have higher raptor incidents.

There is no particular area of the country that is more or less susceptible to avian and bat fatalities than another. All wind sites are intimately connected to their immediate wildlife inhabitants and feeding grounds. Data varies significantly among facilities and regions.

Much has been and is currently being learned yearly about the interference of wind developments on their wildlife neighbors. Many studies show some interesting consistencies in the effort to diminish avian and bat mortality rates. Key conclusions for what is known about land-based wind sites are most often discussed in a three-tier approach of understanding:

- Conclusions supported by peer-reviewed studies—Whereby there is a broad consensus among industry experts. Two types of impacts, *direct mortality* and *indirect impact* have been documented at existing wind developments.

  Direct mortality results from avian collisions with the wind tower itself and the turbine blades as well as **barotrauma**, serious or fatal internal damage resulting from extreme and rapid pressure changes. Migratory songbirds top the list of fatal encounters, as they are the most numerous birds in North America. Their flight patterns are usually above the wind turbine swept area, but fatalities typically occur at stopover sites adjacent to wind facilities, which interfere with takeoff and landings. Hence spring and fall are peak fatality seasons.

  Bats migrate, too, and fatality rates peak at wind sites in the late summer and early fall during the bats' migration. The three migratory bats reportedly most affected are the hoary bat, the eastern red bat, and the silver-haired bat. Currently these three species are not classified as threatened or endangered, but concerns are being heard as more and more wind facilities are in the planning stages and may someday soon encroach into migratory flight paths. Indirect impacts result when birds and bats avoid the wind site area. This leads to what amounts to disruption of familiar habitat and puts nesting and breeding grounds "out of bounds" for the birds and bats. The development of a wind farm can lead to the loss of flora and fauna. Raptors are known to frequent ridge tops, upwind sides of slopes and canyons where wind currents prevail for favorable hunting. Disruption of these natural currents hinders and drives away not only raptors but all migratory avian species. Avoidance and, in extreme cases, abandonment of these essential feeding and breeding grounds become a critical concern in the siting of a wind development.

- **Indeterminate findings**—At this tier of research, field studies and findings are less well understood and no certain conclusion can be drawn due to limited evidence, contrary conclusions, or sometimes even disagreements within the team that conducted the research.

Commercial towers in recent years have tended to be monopole, or single pole, compared with the lattice support towers of previous years. Moving away from the lattice style design eliminates raptor perching sites, thus lowering fatalities. This transition coincides with other technologies and siting practices so that improvements are hard to ascribe for certain to any one factor.

Waterbird and waterfowl interaction at wind towers has had limited study, but field reports indicate low fatality rates. Grassland and shrub-steppe environments, however, show signs of concern. Species of birds such as grouse and prairie chickens inhabiting grassland domains are showing signs of distress **FIGURE 8-6**. A continuous, unfragmented habitat is critical for their survival. Many of the grassland species are known to avoid nesting close to roads, power-line poles, trees, or any domestic foundation or platform. Just what this means for wind tower foundations has yet to be defined.

- **Findings still in the hypothesis stage**—At this stage little is known, and questions remain. Tentative conclusions are suggested due to lack of current information and data gaps. Several demographic areas where little is known and still pending are:

  - **Agricultural vs. forested habitats**—Field studies show signs of lower migratory songbird and bat fatality rates in agricultural landscapes than

FIGURE 8-6 A collaborative research effort initiated in 2006 established the effects of wind developments on prairie chickens in the Midwest.

with wind developments in forested habitats. It is unclear why this is happening but some researchers have suggested fewer songbirds present in agricultural areas and fewer field studies in some landscapes than others; neither scenario, if correct, points to a conclusive comparison.

- Tower height—Taller towers signify longer, wider turbine blades, giving a larger swept area and therefore more overlap of the migratory flight path. Wider and longer turbine blades also produce more blade-tip vortices and blade-wake turbulences. The overall effect on avian and bat populations is still uncertain.

- Motion smear—A study is underway involving two methods of making turbine blades more visible to migratory birds: marking the blades with a color pattern or using an ultraviolet (UV) paint to coat the whole blade. The effectiveness of either method is still to be determined.

- Barotrauma—While the majority of fatalities are thought to happen by direct collision, there is the phenomenon of barotrauma to consider. The rapid and extreme pressure changes due to blade vortices and turbulence is enough to cause internal injury to bats. Researchers have not determined the rate at which this is occurring. Some internal injuries to bats happen immediately, while in other cases, animals are injured but fly away and succumb away from the wind site. Current methods of carcass counting for barotrauma mortality rates do not account for these off-site deaths, so the estimation of bat fatalities is uncertain.

- Nocturnal—The use of night-vision goggles and thermal imaging field studies show that not only are nocturnal birds and bats attracted to the rotating turbine blades but they're actually tracking them by following their path FIGURE 8-7. Insect activity in the higher atmosphere above the turbine and seasonal weather are main contributors to the nighttime incidence of avian and bat fatalities.

After years of study and data collection, numerous steps are being taken to lessen the impact of wind development on avian and bat populations. Researchers, developers, and regulators are seeing improved mortality rates with the following:

- **Cut-in speed**—Altering speed from 7–10 mph to 11 mph is showing signs of reducing bat turbine related deaths by as much as 53 percent to 87 percent. This method of deployment is proving to be quite effective, especially during bat migratory season from late July to mid-October.

- Curtailed speed—Turbines realize 72 percent fewer fatalities when at or near full operational speed. Bats are most susceptible to fatality on both low-wind nights and when turbine blades are operating at maximum rpm. Modest rpm reduction during predictable hours of bat activity reduces fatalities considerably. The estimated cost of curtailing turbines from

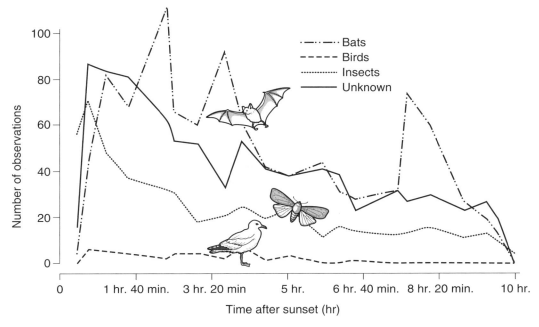

FIGURE 8-7 Charting of nighttime activity levels of bats, birds, and insects via infrared cameras.
Data from www.batsandwind.org/pdf/JWM_M&M.pdf, p. 5

one-half hour before sunset to one-half hour after sunrise resulted in 3 percent lost power output but only 0.3 percent of total annual power output.
- **Deterring devices**—Devices, such as high amplitude sonar jamming sound, are showing promise as a means of reducing bat fatalities.

After years of research, it has been found that the adherence to federal aviation regulations requiring FAA lights be installed on the towerhead of every wind tower has had no correlation with avian and bat fatality rates even though increased insect activity is evident.

Visual Impacts

Today's commercial wind towers are reaching heights well over 200 feet, giving them visual prominence. Public opinion differs in terms of the visual impact a wind development imposes on an area. Some people look up at a ridge of wind turbines or a valley overlaid in turbine towers and see movable art—graceful symbols of economic progress, icons of modern technology. Others may simply see them as eyesores. Either way one looks at them, a wind farm has a large visual impact and the conflicting perceptions require addressing.

Any effort to hide or blur a wind development's whereabouts becomes doomed from the beginning as wind towers exceed the scale of all other elements on the skyline. Their encroachment on the natural or rural landscape requires a visual balance between the elements of the landscape on one hand, and functional aesthetics on the other. The visibility footprint of a wind farm is a top priority for industrial wind designers when they are considering the layout of towers. Designers are now looking at the idea of creating tower clusters rather than using the string approach. This layout concept gives a more intimate grouping sensibility rather than a lined-up array of towers marching down a ridgeline or scattered effect across a valley floor and other pastoral settings. Each grouping highlights an individual sense of placement rather than the massive rigidly placed towers of earlier days.

It is crucial that designers strive for uniformity with respect to the tower design of height and frame structure, rotational speed, color, and rotor diameter among each cluster of towers and across the development site. This continuity of design allows for a perceived sense of rhythm and coherence of the overall visual scope and image. Computer-aided design tools are used in the evaluation phase of development to project the viable footprint before site erection begins, which is critical in promoting a wind site to local contingencies. This is also helpful in facilitating how and where shadow flicker will show up on a proposed development.

Shadow flicker is the term for the shadows created by the motion of the turbine blades. The shadow that a blade casts moves rapidly across the ground and nearby structures. Shadow flicker, while not a direct visual impact, is very important to consider when planning a wind facility project. Although these shadows occur only when the sun is at low angles, they are known to be very unpleasant to occupants of the buildings the shadows move across.

Noise

Wind turbine noise is generated from two types of sources: aerodynamic and mechanical. It is the combination of both that results in turbine noise. Aerodynamic noise is generated by the rotation of the turbine blades through the air and varies according to the blade tip speed in relation to the wind speed. Mechanical noise is generated by the turbine's internal mechanical components such as the gearbox, drive shaft, generator, and rotor assembly rotational movement. On commercial turbines the nacelle is insulated for noise abatement but may have discernible tones, making this type of noise somewhat more noticeable and irritating.

Different turbine models, along with variable wind speed, can create noise that may seem like a buzzing, a whooshing, pulsing, or even sizzling sound. Several turbines located in close proximity of each other can combine their aerodynamic noise to create an oscillating or thumping "wa-wa" sound effect.

Most often the noise radiates perpendicular to the blades' rotation and will change direction depending on the wind and where the rotor assembly faces on any particular day.

Blade pitch and blade speed also contribute an overall broadband frequency from 20–3,600 Hz. Some turbines produce a higher percentage of low frequency sounds at low wind speed than at high wind speeds. Also, any accumulation of debris on the turbine blade will cause a frequency sound. A public utility company using wind-generated power requires the turbine to be compatible with the grid transmission making it necessary for the turbine to keep its blades rotating as constantly as possible. This requires the adjustment of the blade pitch, thus changing the noise level and frequency.

Today's wind towers are designed with noise abatement in mind. Sound is measured by its intensity in decibels (dB). You can hear a frequency pitch typically around 40 to 50 decibels if you stand anywhere within 1,000 feet of a wind tower. It's the equivalent to a kitchen refrigerator running.

Studies have shown that wind turbines generate on the average 40 to 50 decibels. In comparison to other common noise-generating activities this is fairly low and well below the 85 decibels considered harmful when constant. Most often it is very difficult to distinguish the reported turbine noise from nearby disturbances such as vehicle traffic, overhead air traffic, or general community living conditions **FIGURE 8-8** . Even the sound of ground wind across the landscape becomes an **ambient** noise referred to as *masking* and thereby washing out the turbine noise.

Electromagnetic Interference

With the increasing magnitude of commercial wind turbines comes the concern of electromagnetic interference (EMI). The broadcasting of electromagnetic waves, commonly referred to as a radar system, is used as a technological means for tracking the range, altitude, direction, or speed of both moving and fixed objects such as aircraft, ships, spacecraft, motor vehicles, even weather formations and terrain. These waves require a direct line-of-sight in order for detection to take place. With the advancement of commercial wind power generation, these tracking systems are showing signs of interference.

Radar

Radar systems utilize electromagnetic waves in the radio frequency of 3 MHz to around 100 GHz. They consist mainly of a transmitter and a receiver. The transmitter emits an electromagnetic wave (radar beam) across the atmosphere hitting an object, usually a specified target, whereby the radar beam will reflect back to the radar receiver giving direct feedback for change of direction and speed. Electromagnetic interference happens when a degree of difficulty is encountered either by difficulty in processing the reflection or when signal variability occurs.

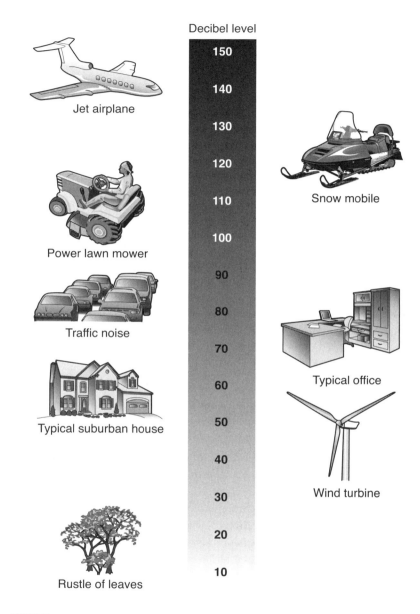

FIGURE 8-8 Comparison of common sound levels.
Data from AURI

Commercial MW wind turbines cause radar beam interference (EMI) in three ways:

- **Near field effects**—When electromagnetic fields are emitted by the generator and switching components in the wind turbine nacelle, they become additional non-transmitter transmitted radar signals that can also be detected at the radar system receiver.

■ **Diffraction effects**—Occur because the tower itself can absorb the radar signal being transmitted. The magnitude of diffraction depends on the number and location of turbines in close proximity of each other. Also, the presence of numerous turbines within a limited area produces a shadowing or background signal, inhibiting the radar to distinguish aircraft from that clutter.

■ **Reflection/scattering**—Occurs when the tower turbine either reflects or obstructs the transmitted signal by its position in the direct line-of-sight to the intended target between the transmitter and receiver. The rotating towerhead blades receive a transmitted signal but transmit a scattered signal back. The receiver may pick up two signals simultaneously or a scattered delayed signal (out of phase) that is distorted to the primary sent signal. Blade tip speed often falls within the same range as aircraft and will appear to a radar station as a moving target of significant size.

Multi-bounce is another term used to describe a radar wave first reflected by a wind turbine blade, then the turbine tower, and once again by the blade before returning to the radar receiver. The rotation of turbine blades can also create a cone of silence directly above the turbines. Blade rotation can lead to the creation of a blackout zone, where planes can disappear from radar entirely. Clusters of turbines can also imitate storm activity on weather radar systems, creating a false positive reading.

Wind developments located in the line-of-sight of US military or commercial airport radar facilities pose a potential threat to radar performing correctly, which makes it difficult for air traffic controllers to receive and give accurate information.

Safety Considerations

A wind development with its individual towers may not seem like an industrial facility, but it is, and safety considerations must be taken into account. However, the wind industry comes with some potential hazards particular to its area of operation and these exceptions require due diligence if a continued track record of safety is to be maintained.

Blade Throw

Blade throw has been known to happen on small-scale, less robust residential and medium-size commercial installations. These installations employ towers less than 100 feet tall. Large commercial MW towers installed at commercial sites for power transmission to the national power grid are proving out to be quite reliable. On these commercial turbines, blade throw is virtually nonexistent. Steady monitoring of turbine components by control systems and sensors detect any deviation to normal operating patterns and deploy shutdown measure immediately. Modern large MW wind turbines with state-of-the-art components have hundreds of thousands of hours of operation without incident.

Falling Ice

Wind turbine blades can accumulate ice under certain atmospheric conditions, such as near freezing temperatures combined with high relative humidity. Also, freezing rain or sleet will cause ice buildup on the surface of a blade. At some point the ice becomes too heavy and falling ice, or shedding, can occur. If the turbine is rotating, a phenomenon called *ice throw* occurs due to gravity and centrifugal forces propelling sheets of ice or ice fragments some distance from the turbine.

Today's MW turbines, with turbine blade rotational heights from the ground to their blade tip upwards of over 400 feet, can easily reach low-lying cloud cover. At this height, super-cooled rain in the cool seasons can begin creating ice buildup starting on the leading edge. Evidence of ice crystals forming on ground cover underneath the turbine, or on the grass around and in line with the rotation of the turbine blades, is a good indication the blades are experiencing ice accumulation. In such a situation, the area should be avoided. Adequate setbacks are employed between each turbine and nearby residences to deter the risk of injury or property damage. Also automatic turbine shutdown controls are deployed when ice buildup conditions on the turbine blades are detected.

Tower Climbing—Slips and Falls

Climbing an average commercial wind tower at a height of 265 feet is like climbing the face of a 20-story building. Wind tower technicians climb almost daily to the turbine towerhead to handle routine and preventive maintenance. There are stairs inside the tubular tower for this purpose, but once the technician is up in the nacelle, he or she must step outside for visual inspections and other routine maintenance.

Wind turbine maintenance presents very specific challenges. The towers can be slippery in wet weather, and tower climbers most often must work in very high winds. This is after climbing a 300-foot ladder wearing heavy work clothing **FIGURE 8-9** and carrying tools and other equipment.

In the event of a slip or fall, it is imperative that a worker be reached immediately with rescue support. Several conditions inherent to working suspended off the ground can set in very quickly. Suspension trauma can occur within 7 to 20 minutes. Because the safety harness restricts blood flow to a person's extremities a condition called "harness hang syndrome" begins. Symptoms that first appear can seem flu-like such as excess sweating, nausea, dizziness, and hot flashes. If the situation continues, brain function deteriorates, followed by difficulty in breathing, elevated heart rate, cardiac arrhythmias, and abrupt increase in blood pressure. Unconsciousness and even death can occur next, which is why it's critical to employ the "two-person rule." Technicians should always work and perform any maintenance with a partner.

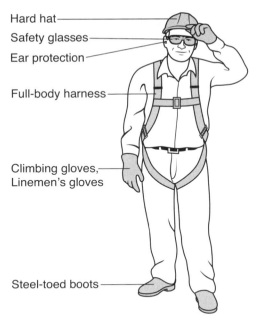

Hard hat
Safety glasses
Ear protection

Full-body harness

Climbing gloves,
Linemen's gloves

Steel-toed boots

FIGURE 8–9 Tower technician in full safety gear.

Adapted from Plantier, K., & Smith, K. M. (2009). *Electromechanical Principles of Wind Turbines For Wind Energy Technicians, 1st ed.* Waco, TX: TSTC Publishing.

THE CENTER FOR ENERGY EFFICIENCY AND RENEWABLE ENERGY (CEERE) AT THE UNIVERSITY OF MASSACHUSETTS, AMHERST

The Center for Energy Efficiency and Renewable Energy (CEERE) represents a university-based program providing technological development and knowledge transfer. Under the auspices of the University of Massachusetts Wind Energy Center, the CEERE program is considered the nation's foremost graduate curriculum, addressing wind energy research nationally and internationally since 1972.

Located at the University of Massachusetts, Amherst, CEERE has capabilities that extend beyond the classroom giving rise to technological and economic solutions to environmental issues around renewable energy production. The Wind Energy Center provides research, training, and educational experiences for graduate and undergraduate engineers and scientists. Central to CEERE's mission is its offering of research programs in collaboration with state, federal, and private partners. Essential to the curriculum are programs that offer resources for long-term research initiatives and spinoffs for contract opportunities.

(Continues)

THE CENTER FOR ENERGY EFFICIENCY AND RENEWABLE ENERGY (CEERE) AT THE UNIVERSITY OF MASSACHUSETTS, AMHERST (Continued)

In the interest of achieving a sustainable future, CEERE's mission statement has at its core the exploration, development, and promotion of economical solutions surrounding the complex relationship between renewal energy and the environment. Research, educational programs, policy setting, planning, and maximized knowledge transfer are all part of the purpose and success of the program. The center and subsequent programs are considered the major hub for federal and state energy programs.

CHAPTER SUMMARY

Wind energy development is here to stay, growing exponentially every year. Its impact on our environment becomes more and more a major concern, as each facility has the potential to reduce, fragment, and degrade its immediate surroundings. Reducing the risk of habitat losses and integrating these industrial facilities into our local communities becomes the challenge. Numerous environmental concerns, if not addressed properly, can delay the building of a wind power facility.

Several key species, such as types of birds and bats, are showing signs of adaptation to wind farms, but the goal of reducing the threats to these creatures is on the mind of every civil servant, designer, manufacturer, and developer in the wind energy field. Environmental stewardship programs with research and data collection address these at-risk species. Scientists hope to discover what it takes to help wildlife survive and thrive around wind farms.

KEY CONCEPTS AND TERMS

Ambient

Avian

Barotrauma

Cut-in speed

Flora and fauna

Footprint

Raptor

CHAPTER ASSESSMENT: ENVIRONMENTAL IMPACT OF WIND SYSTEMS

1. _____ to _____ percent of wind projects are delayed due to concern over adverse environmental impacts.

2. The overall magnitude of land required to develop a wind energy facility can be sizable in comparison to the actual footprint of the tower foundations, roads, and site infrastructure.
   - ❏ A. True
   - ❏ B. False

3. Which of these causes higher mortality among birds than do wind turbines?
   - ❏ A. Buildings/windows
   - ❏ B. Vehicles
   - ❏ C. House cats
   - ❏ D. Pesticides
   - ❏ E. All of the above

4. The counting of animal _____ is the most-used methodology for estimating wildlife fatalities.

5. Population decline by many bird and bat species is due to:
   - ❑ A. habitat loss.
   - ❑ B. fragmentation.
   - ❑ C. other invasive species.
   - ❑ D. human interference.
   - ❑ E. All of the above

6. Wind tower turbine blades are the main contributor to avian fatalities.
   - ❑ A. True
   - ❑ B. False

7. The study of wind developments on wildlife habitat are most often discussed in a three-tier approach. Which three are considered to be part of this tiered approach? (Select three.)
   - ❑ A. Still under advisement
   - ❑ B. Conclusions supported by peer-reviewed studies
   - ❑ C. Indeterminate findings
   - ❑ D. Findings still in the hypothesis stage

8. _____ employs two methods of making turbine blades more visible to migratory birds by marking the blades with a color pattern or using an ultraviolet (UV) paint to coat the whole blade.

9. What has proven to be the best method of reducing bat-turbine related deaths during the bat migratory season from late July to mid-October?
   - ❑ A. Turbine cut-in speed
   - ❑ B. Curtailing of the turbine
   - ❑ C. FAA lights
   - ❑ D. Deterring devices

10. Shadow _____ is the term for shadows created by the motion of the turbine blades, which cast a shadow that moves rapidly across the ground and nearby structures.

11. Wind turbine noise is generated from what two types of sources? (Select two.)
    - ❑ A. Ambient
    - ❑ B. Aerodynamic
    - ❑ C. Sonar
    - ❑ D. Mechanical

12. Adjustment of the turbine blade pitch changes the frequency and emitted sound level (noise) of the turbine.
    - ❑ A. True
    - ❑ B. False

13. A wind turbine noise levels have been tested to be between 35 to 45 decibels, which is louder than:
- ❏ **A.** an airplane flying overhead.
- ❏ **B.** a whisper.
- ❏ **C.** inside of a car while driving.
- ❏ **D.** a pneumatic drill.
- ❏ **E.** industrial sounds.

14. Barotrauma is the phenomena of rapid and extreme pressure changes due to blade vortices and turbulence. It is not enough to cause serious injury or death to bats.
- ❏ **A.** True
- ❏ **B.** False

15. When wind turbine blades accumulate ice under certain atmospheric conditions such as near freezing temperatures combined with high relative humidity, it is called:
- ❏ **A.** shedding.
- ❏ **B.** ice throw.
- ❏ **C.** ice buildup.
- ❏ **D.** falling ice.
- ❏ **E.** All of the above

Economic Prospects of Large Turbines

THE ECONOMIC PROSPECTS of large wind turbines can often determine whether a project will progress to full energy production. Because the profitability of a wind development cannot be determined by machine costs alone, it's worth it to address the influence of site issues, operations and maintenance expense, and infrastructure costs on the overall feasibility of a turnkey wind facility.

The financial success of a wind development depends on the comparative amount of electricity sold to the utility company or end user versus the cost required to find, extract, process, manage, and deliver this energy commodity in a usable form for consumption. For any investor or developer, a return on investment (ROI) calculation is the financial standard used to project a venture capital project to be profitable.

There are many methodologies for the determination of a ROI. The disadvantage of a conventional ROI calculation for the energy industry is that it does not account for the sustainability of energy resources and the size of the carbon footprint each type of energy fuel leaves. Hence, the energy industry has modified the ROI calculation to reflect an energy return on investment (EROI) calculation.

This chapter discusses how much fossil fuel is required in the development of a wind facility, such as the diesel used in transporting components to the wind site, or the amount of fossil fuels used to manufacture components, and many other peripheral factors. The EROI calculation is the last area of concern before the economic prospects of a wind development are deemed viable. If the investment does not have a high enough or positive EROI, or there are other opportunities with a higher percentage of EROI, then the investment is not considered to be energy efficient.

Chapter Topics

This chapter covers the following topics and concepts:

- Introduction to the concept of EROI and the definition of ROI in comparison to EROI
- Energy production costs in relation to net EROI
- Evaluating electricity generation costs in a wind energy development
- The integration of renewable systems

Chapter Goals

When you complete this chapter, you will be able to:

- Identify the energy costs used in an EROI ratio and relate how the additional developmental costs of site, O&M, and financing effect the overall EROI
- Discuss the feasibility of wind energy production in comparison with conventional energy production
- Relate the comparative elements of an EROI ratio calculation from an energy-production costs point of view to that of energy-generation costs

EROI Introduction

The energy industry, with its special criteria, has modified the concept of the conventional ROI calculations to a higher level of expectation. The energy industry decided to keep in mind the desire to lower our carbon-producing energy use and the corresponding reduction of our carbon footprint on planet Earth.

EROI calculations take into account the additional fossil fuel and electricity required for the manufacture of components, transportation of components to construction sites, the construction of the facility itself, operation and maintenance over the lifetime of the facility, overhead and grid connection costs, and so forth. They represent the external fuels required over and above the proposed energy generation being considered. This is the measure of the additional energy that must be consumed in order to generate the proposed energy (i.e., wind energy) to give a *net energy* supplied. This is a ratio calculation and can be stated in equation form as:

$$EROI = \frac{\text{Quantity of energy supplied to end user}}{\text{Quantity of energy used in the process of production}}$$

The numerator in this equation, "Quantity of energy supplied to end user," is an easy measure to identify, but not as easy to quantify. It is the kW per hour rating of the entire wind facility given as an annual energy output figure to the utility company. This determination gets tricky when you consider the uncontrolled, intermittent nature of wind. This variability affects the EROI value with its mismatch to the utility's energy demands, so an evaluation is required.

It is the denominator figure "Quantity of energy used in the process of production" in the above equation that can be a rather nebulous figure and open for debate. Just how extensive into the energy-cost requirement this analysis should go, is an ongoing debate. The analysis can go beyond the pure physical measure taking into consideration peripheral energy costs such as R&D, transportation fuels, costs of manufacturing plant electricity, the energy costs of environmental study and government regulation, and so on. The more extensively the developers explore these costs, the higher this figure becomes—reducing the EROI and rendering the ratio figure useless if all sectors of the industry are not following the same procedure or comparison.

When making the calculation, you'll want the numerator (energy produced) of the EROI fraction to be greater than the denominator (energy expended). Where conventional ROIs are given in percentages of profit, the EROI is given as a ratio and further calculated giving a resultant whole number for final analysis comparisons. For example, an EROI calculation would be the amount of fuel (energy) it takes to construct, build, drill, road, process, and upgrade a specific fuel type into a socially useful form. This total **energy cost**—the amount of combustible fuel (fossil, nuclear, gas, etc.) required to support the production of a service or product—would then be divided into the annual amount of electricity sold.

Development of a **turnkey,** or production-ready, wind energy project requires looking at the total overall energy expenditures of the wind facility. This additional and final net-energy analysis, or EROI, is of major importance and can be significant enough to halt the progress of a development.

The total EROI of a wind facility can be broken down into two main areas of consideration, each having its own criteria in support of the final overall EROI for a wind facility. The first consideration is energy production costs. This is the ratio comparison of wind energy capture capabilities of the tower and wind turbines to the energy costs of manufacturing the tower and wind turbine.

The second consideration is the electricity generating costs. This is a ratio comparison whereby the produced electricity of a wind facility is compared with the energy costs of running, maintaining, and operating the facility. Both are looked at in reference to the added energy costs (fossil fuels for transportation, electricity usage in manufacturing, etc.) that represent a carbon footprint on our Earth over and above the renewable and sustainable energy being produced.

Also, as all investors and developers know, it is not just how much energy profit you receive in return for your investment, but also how long it takes for the payback. The average life expectancy of a wind turbine is 20 years, and you will use that figure for the purpose of this next discussion.

Energy Production Costs in Relation to Net EROI

Energy production costs in relation to a net EROI analysis are concerned mainly with the tower and wind turbine. Tower height and turbine size are directly related; the higher the tower profile, the greater the rotor diameter, allowing for more energy capture. However, when turbine size increases, the respective machine energy costs occur at a faster rate than the rate of energy capture. An EROI analysis for tower and turbine size comparison is warranted whereby the specific EROI ratio for energy production costs is demonstrated in the following equation:

$$EROI = \frac{\text{Cumulative energy produced}}{\text{Cumulative cost-of-construction energy}}$$

This is a simplified statement, as many calculations go into the cumulative aspect of the above equation and subsequent ratio calculation. This is where the manufacturer and designers come into play. It is their job to design larger turbines and taller towers that are less expensive to build and install, with higher efficiency and less energy cost to operate. Generally speaking, an increase in the power rating of a wind turbine means an increase in the EROI, but a high-efficiency rated turbine with a small rotor diameter does not accomplish an overall higher EROI.

The first means of raising the overall EROI of a wind tower is to also raise the tower height. This will accomplish two things to achieve more efficient wind energy capture. First, it gets the wind turbine higher up in the atmosphere, where the winds are more predictable. Wind turbines with a taller profile can extract wind energy from the higher, steadier wind velocities that exist at greater heights. These velocities are the result of less surface roughness of the surrounding land mass. Often the height and type of vegetation and buildings tend to reduce wind velocity near the ground. Also, an open flat terrain allows for the wind speeds to increase relatively fast with tower height.

This wind speed increase with the increase in tower height allows for better wind speed calculations, which leads to greater and more stable wind tip speed calculations. This in turn produces better annual power ratings to the utility companies. The combination of increased wind speed and improved constancy of speed produces a higher *cumulative energy produced* number for the EROI ratio calculation as stated above.

The second area of improvement that comes with a higher profile tower is an increase in the rotor diameter. Once the wind turbine is higher off the ground, longer, larger rotor blades are then appropriate. A larger rotor diameter determines the swept area of the rotor assembly, thereby giving a larger area for wind capture and greater energy production. This is probably the single most important factor for the potential to capture and generate more wind power thereby increasing the EROI of a wind turbine. A larger rotor diameter requires a greater initial investment of materials (i.e., energy production usage) but the increase in EROI more than compensates for this.

The *cube rule* is another reason the megawatt turbines have a larger EROI. The cube rule states as the wind speed doubles, the power of the wind increases eight times. The new taller megawatt towers take advantage of this rule by allowing the wind turbine to extract wind energy from the higher velocity winds that exist at greater heights. The initial money invested, or **capital** costs, of machinery and construction materials will be more, but with the new megawatt (MW) range of operations also come technological advances improving the overall power output, which more than compensates for the higher initial energy cost of construction.

The number of rotor assembly blades on a wind turbine also plays a role in its EROI numbers. A two-blade rotor assembly will have different numbers from a three-blade turbine. A two-blade turbine will have less operation and maintenance (O&M) expense as compared to a three-blade machine; however, a two-blade machine does not have the same efficiency; this results in a lower power output. Two blades do not have the same overall capturing capability as three. This makes the three-blade turbine, from the perspective of energy production and energy cost comparative calculations (EROI), the overall choice.

One last area of analysis for the EROI of a wind project is the tower placement within the wind development. The EROI is influenced by the location of each tower within the development site. A current trend is leading away from the traditional string arrangement of towers. Wind interaction and related disruptive turbulence between towers in a string pattern have been documented to reduce energy capture between 3 percent and 10 percent, depending upon the number of towers in the array. A current trend of placing towers in clusters with each tower placed for maximum wind resource gives a better EROI.

The current trend in the wind industry of increased tower and turbine size into the megawatt range has a direct effect on the total project's EROI. This calculation has become the standard benchmark throughout the energy industry. Wind power in comparison to other conventional sources of electricity generation compares favorably, as shown in **FIGURE 9-1**. It also computes to better energy return when compared with current-day renewable and sustainable energy sources such as solar and geothermal.

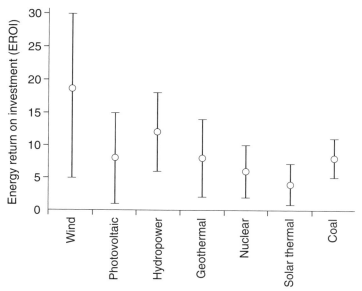

FIGURE 9-1 Comparison of wind energy to other power system EROI.
Data from http://www.eoearth.org/article/Energy_return_on_investment_(EROI)_for_wind_energy

Electricity Generation Costs

Electricity generation costs entail comparison of the ex-works components of a wind development against their fuel consumption both during construction and the life expectancy of the facility. Ex-works elements encompass the infrastructure of the facility, such as road building and maintenance, overhead power lines, electrical equipment on the ground, and an electrical collection system including all support structures and services. Support structures are the facility substation or substations and maintenance buildings. Support services mainly manage the operation and maintenance not only during construction but for the life of the facility. Designers should also consider possible grid connection and decommissioning of the facility at the end of its life expectancy.

Wind energy facilities have external costs as well. Even though they are hard to assess in monetary and energy-cost terms, they do require consideration. Researchers are just beginning to assess the environmental, wildlife, and human energy costs. As previously discussed, the downstream effect of wind facilities on their immediate environment has consequences. The exact energy-cost amount is still undefined but there is indication that field irrigation and crops may require extra measures to ensure their productivity levels are maintained. Thus, electricity consumption for irrigation over and above the seasonal norm will be tallied as an energy cost related to wind technology.

The electricity consumption for additional irrigation and the capital cost of machinery as discussed previously is what defines the total EROI of a wind development and the bottom-line criterion for judging the success or failure of a facility as a viable renewable and sustainable energy source. This comparison was demonstrated previously in equation form as the EROI calculation in a ratio format. FIGURE 9-2 illustrates this same concept to help show the relation of energy costs extending over the life expectancy of a standard wind facility.

This graph allows you to see the cumulative energy expenditures over the life expectancy of a wind facility to include decommissioning after the facility warrants itself beyond cost-effective repair. This can also be demonstrated in the following ratio equation:

$$\text{EROI} = \frac{\text{Cumulative energy production}}{\text{Cumulative energy costs}}$$

The comparison of conventional fossil fuels such as coal, oil, and gas to renewable energy shows an overall picture in which fossil fuels have a much shorter ROI; however, the EROI of these fuels is much debated. The wind industry seems to be gaining ground in its attempt to produce more electrical energy at a better EROI.

A compelling illustration of the growing size and scope of modern wind projects is the Alta-Oak Creek Mojave Project at Tehachapi Pass, California. This wind

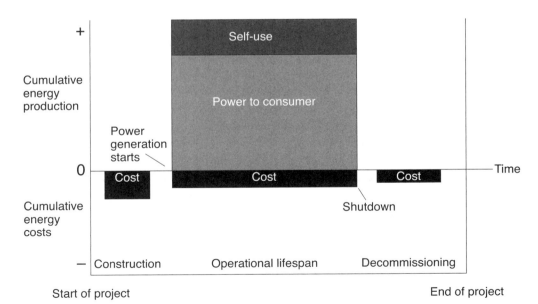

FIGURE 9-2 Cumulative energy costs and production over a facility's life cycle.
Data from The Encyclopedia of Earth

project encompasses 320 wind turbines installed over a 9,000-acre site producing 800 MW of power. The project was started in 2010 and is expected to reduce carbon dioxide emissions by more than 52 million metric tons, or the equivalent of taking 446,000 cars off the road. The Alta-Oak Creek Mojave Project accomplishment is due to a 25-year power purchase agreement whereby clean-air wind power will be used in place of fossil-fuel power generation.

A designer also needs to make important assumptions regarding operating characteristics. These include selection of the power rating of the turbines, their expected lifetime, and the capacity factors. Changes in these assumptions or deviation in the actual operating conditions from the assumed conditions has significant impact on the final EROI results. Also, the intermittent nature of wind plays a role in reducing the EROI value and has an effect on the payback that can be expected. The payback timeline is how many months of generated wind electricity are required to pay off the capital to cover the costs of the wind development. This can take some looking into, as it entails the wind systems integration into a major power grid and must take into account the utility suppliers' future energy predictions.

Renewable Systems Integration

Utility companies increasingly are considering the use of wind-generated power for a more prominent role in their total energy production. The variability and uncertainty of wind production inherent in wind power is a concern. Wind-generated power does not always match up with the variations in electricity load demands of the current conventional system.

The intermittent and cyclic nature of wind relative to operator-controlled energy sources such as coal, gas, or nuclear fuel-generated power plants makes it a hard commodity to resource. In addition, the inability to store wind energy or have on-demand use, as hydroelectric power plants do with their dams and reservoirs, adds to the dilemma. The intermittent nature of wind makes the integration of its electrical power into the national power grid an issue for both the utilities and the wind developers.

From the utility companies' standpoint, the sporadic nature of wind makes it unacceptable as a reserve energy supply that can be drawn upon when needed, not just when available. Surplus wind power during times of high wind capacity cannot always be utilized at that moment by conventional operating processes. When a wind turbine's capture of raw energy is overabundant or greater than the amount of electricity the power grid requires at that time, the wind power is refused by the grid. Energy is being produced but not utilized—or sold. This affects the wind developer's ROI and more specifically the EROI ratio of the facility.

A reverse of this overabundant situation happens when the power grid cannot rely on or predict the availability of steady electrical input from the wind facility due to seasonal lows, or common daily fluctuation in the wind flow pattern.

This mismatch of wind energy to regional energy demands is a major concern to the industry. How to increase the reliability of wind-produced electricity has prompted extensive investigation. A number of programs are in place to envision and promote better wind utilization for the power grids. Many people are working diligently toward increasing the integration and confidence in the reliability of new wind turbine projects **FIGURE 9-3**.

The Wind and Water Power Program addresses these needs by working with current grid operators and planners, regional utility companies, and national regulators. Their goal is to create new strategies allowing for more production of wind energy to be incorporated into the conventional power grid system while maintaining reliable grid operations.

Several areas of study and program assistance are being employed in this ongoing effort to help promote and accomplish better utilization of this intermittent wind energy into the routine power system operations. New strategies look at three main areas of concern:

- Weather forecasting—Plays a very important role in predicting the availability and strength (velocity) of wind energy. Most notably wind forecasting as a subsequent element to weather forecasting that uses state-of-the-art technology for improving accuracy. Along with weather forecasting, evaluation studies promote the best presentation methods of data forecast to systems operators for analysis and use.
- Grid planning—Grid operators, utilities, regulators, and others who address transmission issues are already limiting wind energy development in some areas. This indicates with clear certainty the need for significant new

FIGURE 9-3 Wind energy and power grid integration are increasingly more important as the wind industry grows.
© Lisa S./ShutterStock, Inc.

transmission lines to accommodate the energy that higher towers and larger-capacity wind turbines are capable of supplying. National energy-based transmission planning is required to ensure both consistent and economical energy delivery across a larger geographical area than the immediate region of a wind facility. The sharing and transferring of electrical energy between the smaller regional power grids and the major power grids is a valid option. For example, excess wind energy is drawn into the grid system from high-yield regions and passed along to areas of *energy sink* (areas with low or no energy production). Grid planning can take this into consideration for the placement of transmission lines and substations in the future.

■ Storage—Or on-demand use of wind energy is proving to be a very hot topic for the future of this renewable energy. Every industry is working to develop storage systems large enough to accommodate the grid-ready surplus of energy produced. Designers are considering the following to store surplus energy: electrochemical batteries, capacitors, compressed air, pumped hydroelectric, and magnetic and thermal energy storage.

Wind can be controlled somewhat, as previously discussed, through mechanical means of pitch, feathering, and yawing, but never to the extent that conventional generating facilities (gas, coal, etc.) can be controlled. However, with the advent of more sophisticated monitoring and transmission sharing, wind is gaining ground on these issues. Consequently, wind is being taken more seriously for increased electrical input to the national power grid than ever before.

CHALLENGES TO WIND FARM DEVELOPMENT

Exponential growth in the wind industry brings with it its own set of problems and issues. There are many challenges to wind farm development, as you have learned. There are several major challenges that stand out as a possible hindrance to the evolution of the wind industry including:

■ Megawatt size—Growth of the wind tower and consequently its rotor diameter (blade length) brings a whole set of transportation issues. Rotor blades, many in excess of 160 feet, cannot be shipped over many off-highway roads and bridges that are not built to accommodate such great lengths FIGURE 9-4. Another shipping dilemma is the transfer of nacelles ranging upwards of 50 tons or more. The cost of upgrading or rebuilding these highway structures must be considered as part of the wind development EROI.

CHALLENGES TO WIND FARM DEVELOPMENT (Continued)

FIGURE 9–4 Longer turbine blades give rise to transportation roadblocks.
© iStockphoto.com/arturoli

It is being suggested that the megawatt towers are fast approaching their size limits due to this issue of transportation. Turbine blades are designed whereby all their components must be built in one contiguous piece. Experts feel the length of the blade is reaching its maximum. The bigger turbines may be capable of producing literally more electricity than the local regional utility knows what to do with. Since this energy has no means of being stored, all of the available energy is simply not utilized. In most cases the energy capture is not even allowed, making the whole wind facility redundant.

During times of high wind capacity, it is possible for the wind facility to put more electricity onto the power grid than it needs. In this case, the wind developer suffers monetary losses when the available raw wind energy and subsequent wind power generated is not utilized. Consequently, the facility's profit capacity diminishes during such times, leaving the developer questioning the ROI and also the EROI ratio.

- **Capacity of utility grid transmission lines**—These lines are proving to be outdated and undercapacity. This is significant in the issue of planning for new wind development. In some regions such as Texas, wind electrical output curtailment is employed due to regional trunk lines being already congested and operating at physical capacity, to the point where they cannot take further input. Minnesota and California are examining ways to alleviate transmission overload as more development sites are proposed. These ex-works costs are not always reflected in the EROI comparative analyses, but developers are starting to give much more consideration when the cost of upgrading transmission lines is prohibitive to the wind project.
- **Government subsidies**—Tax breaks, such as the wind energy Production Tax Credit (PTC) in the United States and the so-called feed-in tariffs in other countries, are uncertain for

(Continues)

CHALLENGES TO WIND FARM DEVELOPMENT (Continued)

future projects. The PTC was scheduled to expire on December 31, 2005, but was extended to December 31, 2007, and extended again on February 17, 2009, adding a new incentive. For wind, the PTC was extended an additional two years until the end of 2012. This marks the fourth time the PTC has been extended.

Other incentives such as grants in lieu of tax credits are meant to help those wind developers who weren't so profitable, and showed losses whereby taxes were not applicable. Even though the PTC has been extended by Congress several times, it has also been allowed to "sunset"—expire—three times. This inconsistent nature of the tax credit makes for a significant challenge to wind developers needing long-term planning. Wind projects need more than two years just for meteorological testing, for instance, before they can be moved forward.

CHAPTER SUMMARY

The development of a wind facility is not unlike any other capital business venture where the bottom line is its profitability. ROI, used as a standard measure, has been modified by the energy industry with a higher expectation—EROI calculation. This energy investment calculation sets a higher standard of comparison to include the effects of carbon emissions and the overall carbon footprint left by the exploration and production of energy-related fuels.

Wind energy EROI compares favorably not only with the conventional fossil fuels but also with other renewable resources. New technologies and better weather-predicting capabilities are leading the way for wind energy to be more responsive to the national power grid. Programs are underway to address the electricity market's interconnection impacts, operating strategies, and system planning. This is allowing the wind industry to compete without disadvantage and come closer to meeting its goals for providing more energy for the national economy.

KEY CONCEPTS AND TERMS

Capital
Energy cost
Turnkey

CHAPTER ASSESSMENT: ECONOMIC PROSPECTS OF LARGE TURBINES

1. The return on investment for the energy industry that calculates a higher level of expectation is called:
   - ❏ **A.** return on investment.
   - ❏ **B.** energy cost of investment.
   - ❏ **C.** energy return on investment (EROI).

2. EROI is the ratio of the additional energy consumption that must be expended in order to generate the proposed energy (wind energy in our case) to give a figure for "net energy" supplied.
   - ❏ **A.** True
   - ❏ **B.** False

3. The EROI takes into consideration the desire to lower _____ producing energy and the reduction of our _____ footprint on the planet Earth.

4. The EROI of a wind facility can be broken down into two main areas of concern: (Select two.)
   - ❏ **A.** energy production costs.
   - ❏ **B.** energy machinery costs.
   - ❏ **C.** energy generation costs.
   - ❏ **D.** energy return on investment costs.

5. The cube rule states that as the wind speed doubles, the power of the wind increases _____ times.
   - ❏ **A.** four
   - ❏ **B.** eight
   - ❏ **C.** two
   - ❏ **D.** ten

6. The EROI analysis goes beyond the pure physical measure of electricity, taking into consideration peripheral energy-cost such as:
   - ❏ **A.** R&D.
   - ❏ **B.** costs of manufacturing plant electricity.
   - ❏ **C.** environmental study.
   - ❏ **D.** government regulation energy costs.
   - ❏ **E.** All of the above

7. Energy production costs in relation to a net EROI analysis are concerned mainly with which two things: (Select two.)
   - ❏ **A.** towerhead.
   - ❏ **B.** tower.
   - ❏ **C.** wind turbine.
   - ❏ **D.** nacelle.

8. A high efficiency rated turbine with a small rotor diameter does not accomplish an overall higher EROI.
   - ❏ **A.** True
   - ❏ **B.** False

9. The single most important factor for the potential to capture and generate more wind power thereby increasing the EROI of a wind turbine is:
   - ❏ **A.** generator capacity.
   - ❏ **B.** blade weight.
   - ❏ **C.** rotor diameter.
   - ❏ **D.** tower location.

10. A three-blade turbine does not have the overall capturing capability as that of a two-blade rotor assembly making the two-blade turbine from an energy production and energy cost comparative calculations (EROI) the overall choice.
    - ❏ **A.** True
    - ❏ **B.** False

11. Wind interaction and related disruptive turbulence between towers in a string pattern have been documented to reduce energy capture between _____ and _____ percent depending upon the number of towers in the string.

12. Wind power in comparison to other conventional sources of electricity generation compares favorably. It also computes to better EROI when compared to current day renewable and sustainable energy sources such as solar and geothermal.
- ❏ **A.** True
- ❏ **B.** False

13. The _____ nature of wind relative to operator-controlled energy sources such as coal, gas, or nuclear fuel-generated power plants makes it a hard commodity to resource.

14. In an effort to help promote and accomplish better utilization of intermittent wind energy into the routine power system operations, what three areas of study and program assistance are being employed? (Select three.)
- ❏ **A.** Grid planning
- ❏ **B.** Regulatory easements
- ❏ **C.** Weather forecasting
- ❏ **D.** Storage
- ❏ **E.** Location assessments

15. Exponential growth in the wind industry brings with it its own set of problems and issues. What three challenges stand out as possible hindrances to the evolution of the wind industry? (Select three.)
- ❏ **F.** Megawatt turbine size
- ❏ **G.** Capacity of transmission lines
- ❏ **H.** Government subsidies and tax breaks
- ❏ **I.** Regional regulatory issues
- ❏ **J.** Environmental constraints

Design of Offshore Wind Farms

MANY PEOPLE IN THE WIND INDUSTRY believe the future of wind-generated power lies offshore. This chapter will focus on the future of wind energy by looking at current industry theories and developments in conjunction with the design of offshore wind farms.

Europe is the current world leader in offshore wind development. Several European countries constructed offshore facilities as early as the 1990s. As of 2010, there were 39 offshore wind farms in the Atlantic waters off such countries as Britain, Belgium, Denmark, Finland, Germany, Ireland, the Netherlands, Norway, and Sweden. Their total operating capacity currently stands at 2.4 GW with plans to reach 100 GW by 2040. The largest offshore European project, as of November 2010, is the Thanet Offshore Wind Farm, developed by Britain with a supply of 300 MW. Denmark's Horns Rev II project is 209 MW.

Canada's province of Ontario is proposing several projects located off the Canadian shore of the Great Lakes. The Trillium Power Wind 1 project will be approximately 12 miles from shore and over 400 MW in projected power production. Canada also has plans to develop a wind project on the Pacific Coast.

The United States is just now beginning the permit process for its first offshore development, called the Cape Wind Project, in Nantucket Sound, south of Cape Cod, Massachusetts. A second proposed development is in the planning stages off the coastline of Delaware. Each development is striving to be the first US offshore project in North America. Other proposed projects are under consideration not only for the East Coast but the Great Lakes and the Pacific Coast as well.

The United States has not pursued offshore development on its own **continental shelf** as diligently as European countries have because it hasn't needed to. The wide-open Midwestern prairies are as highly rated in wind capacity as most offshore locations. The Western US states, with their vast open regions and readily available government land, are a greater land resource than their European counterparts. The United States also has chosen to focus on land-based facilities because cost of construction is much less than that of offshore developments.

Another important factor that hinders US offshore development is ocean depth. The continental shelf off the US shoreline is much deeper and more severe than European coastlines. The United States would have to modify European offshore technology, designing it to address the harsher US conditions. The shared technologies are just now starting to be released that will enable offshore developments in the more severe conditions of the US continental shelf.

Chapter Topics

This chapter covers the following topics and concepts:

- Offshore wind capacity in relation to depth and harshness of the ocean seabed
- Issues with designing offshore wind farms

Chapter Goals

When you complete this chapter, you will be able to:

- Relate why offshore wind development is considered the future for many developers in the wind industry
- Understand why US offshore wind development lags behind European installations of wind facilities
- Relate offshore wind resources to land-based resources and the difference in technologies required of each
- Discuss the issues and technology demands needed for the United States to move forward in offshore wind energy development

Offshore Development

The term **offshore** means "away from the mainland and in the ocean." Understanding offshore development of wind energy requires you to look worldwide and appreciate the magnitude available for such wind resources. Only in the last several years has the technology existed allowing a full global map of wind velocity comparison to be generated. Study of such maps is allowing developers to accelerate the development of renewable wind energy.

Take special note of the scale for wind speed on the following global wind map FIGURE 10-1. Wind speed over land is scaled from 3 to 9 m/s (meters per second) while wind speed over water scales at 5 to 20 m/s, showing the wide variance in land to water wind speeds. However, note how much of the US Midwest rates high on the "Wind speed over land" scale (see Figure 10-1), which computes to higher wind speeds than the majority of coastal wind speeds for almost every country.

On a US wind map, you can see in more detail how its coastal waters rank high in wind capacity and appear to have huge energy resources offshore on the East and West Coasts as well as the Great Lakes FIGURE 10-2. It is estimated that these oceanic wind resources have the potential to provide up to 900 GW of generating capacity to the national power grid. However, only 10 percent or roughly 90 GW of this potential wind power is estimated to be over waters less than 100 feet deep. These depths are comparable to existing offshore facilities developed in Europe's shallow waters, where proven technologies already exist.

The remaining 90 percent of estimated offshore wind potential is in harsher conditions. These more severe conditions present an obstacle for US

Wind speed over land

3   6   9

FIGURE 10-1 Global wind map showing land-based wind speeds.

Adapted from http://www.renewableenergyworld.com/rea/partner/3tier/images/first-ever-global-wind-map-promises-to-accelerate-development-of-renewable-energy-projects-51741/47084

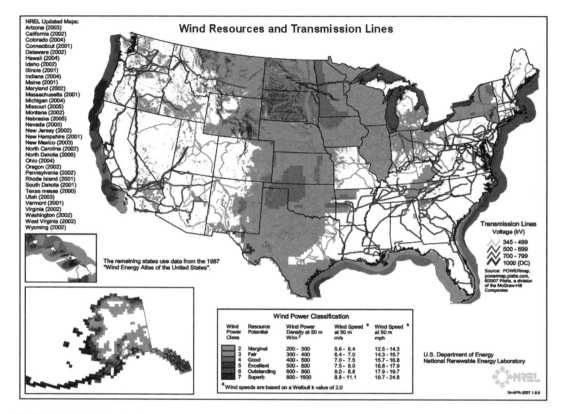

FIGURE 10-2 Wind velocity and occurrence map of the United States.
Courtesy of NREL

developers due to the greater depths, steeper seafloor incline, higher wind velocities, and greater wave action. Researchers and developers need to focus on creating new technologies for tower, foundation, and blade design to address these
conditions.

Several factors give rise to the great interest in offshore wind development:

- **More robust wind environment**—Means higher yield for the dollar invested
  and a better possibility of meeting the US goal of 20 percent wind generated
  energy by 2030.

One of the main driving forces of land-based wind facilities is the
topographic effect as it relates to man-made and natural features of the
earth's surface. With no discernable topographic features offshore, only the
sea's surface roughness comes into play setting up a complex relationship.
On land, the topographic relationship to Earth's surface roughness is fixed
by elements such as trees, hills, buildings, etc. Offshore surface roughness is

dependent on the state of the sea, increasing with wave action, which in turn is influenced by wind conditions. A low surface roughness results in lower wind turbulence, reducing the wind turbine mechanical loads and thereby increasing the energy capture as compared to an identical wind turbine on land with an identical mean wind speed.

Offshore winds on the average tend to flow at higher velocities than land-based winds due to less surface roughness, especially in deeper waters where the average wind speed is considerably higher. This allows for higher annual utility power utilization rates due to the cube rule, which states for every incremental increase in wind speed, energy capture increases exponentially by a factor of eight.

Increased offshore wind speeds of only a few miles per hour equate to much higher potential for energy production. For instance, an offshore turbine with an average wind speed of 16 mph will produce 50 percent more power than a site with the same turbine size but only average wind speeds of 14 mph. This directly relates to the effects of the cube rule.

- **Location, location, location**—Much of the offshore potential wind energy is near major populations, also described as energy load centers. These energy-hungry centers are given crucial consideration to the development of wind energy.

  Land-based commercial wind facilities requiring an abundance of wind are most often sited in remote locales. Getting wind power from such facilities to high population (high-energy consumption) load center areas has been challenging and costly.

  Getting power to areas in need over outdated transmission lines or lines that are already operating at capacity requires a massive reorganization of the national transmission grid. This is not the case with oceanic wind resources; once new cabling from the offshore wind site reaches land, it is already near some of the most populated communities and business centers in the United States.

- **Much larger turbines**—These are now being considered, as offshore installations do not have the same transportation constraints as land-based turbines. Transportation of an offshore turbine blade does not involve maneuvering over rural secondary roads. And, once a blade is delivered to a respective seaport, the ocean transport to job site is merely a matter of using transport barges, which are very capable of handling extended lengths.

  Everything about an offshore wind site is bigger and more expansive than a land-based site where a typical tower is maxed out at a tower height of 200 to 265 feet and blades that are 100 to 130 feet long, which equates to a 200–250 foot rotor diameter (swept area). Comparable offshore wind turbines generally start at the top end of this range, with rotor diameters starting at

250 feet increasing to 350 feet. It is also projected that offshore turbines will soon reach 10 MW.

■ Less obtrusive installation—Offshore towers with their apparent size and noise are greatly diminished by distance. Today the overall development of offshore wind energy installed worldwide has been in shallow waters of less than 100 feet and within five miles of the coastline. This is the case mostly off the coasts of Europe, yet they are still far enough away from their respective seashores that people on land have little or no chance of seeing them. The possibility of hearing any blade frequency noise, as discussed with land-based wind facilities, also becomes negligible.

The New England and mid-Atlantic coasts have been identified as having the best US wind potential and also water depths that gradually deepen with distance from their respective shores. The Great Lakes region also shows great promise, with the West Coast and Gulf Coast not yet fully characterized.

Issues With Designing Offshore Wind Farms

Great strides are taking place in the development of innovative wind technology, especially in the area of expanding the wind industry to offshore. Researchers are focused on using technology that lowers the cost of energy produced by offshore wind turbines via lowering capital costs, increasing reliability, decreasing O&M, and increasing energy capture.

The technical needs and design of an offshore wind farm deviate from that of a land-based development because of the demanding tower foundation requirements and the harsher environment they are expected to reside within **FIGURE 10-3**. Modifications are needed to prevent corrosion due to the salt seawater and protection against wave and wind interaction. As technology advances, allowing for siting of towers in deeper water, so does the style of foundation and layout of a wind farm change.

Commercial-scale offshore wind developments may still have many criteria similar to land-based wind facilities but with needed modifications to withstand the ocean and saltwater environmental exposures. Wind and wave action play a major role in the design and development of the wind tower platform. Terminology used when addressing wave action in reference to the design of offshore tower platforms is shown in **FIGURE 10-4**.

Offshore wind towers look interchangeable with land-based towers; however, they have several modifications in design. The tower, in order to cope with ocean wind and wave interactions, must be of greater strength. In fact, all structural components exposed to the environmental elements must be protected against

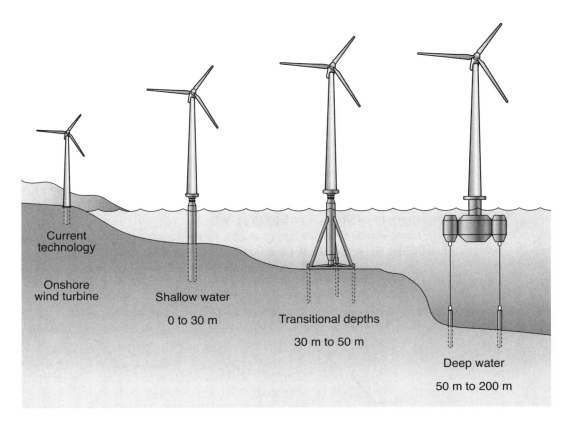

FIGURE 10-3 Evolution of wind tower design into deep-water technology.
Adapted from OffshoreWind.net

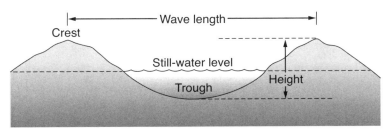

FIGURE 10-4 Terminology used when addressing wave action in reference to the design of offshore tower platforms.
Adapted from NREL

the highly corrosive sea air. A high-grade exterior paint will be used and any exposed fittings will be treated regardless if they are internal or external.

The turbine itself must be protected against corrosion, as must all its internal mechanical components such as internal climate control, automatic greasing systems, and cooling systems. Internal service cranes are employed to minimize servicing. Lightning strikes occur frequently in some locations and installation of lightning protection systems helps minimize risk and damage. Brightly colored paint, such as contrasting yellow, is typically used for the access platform to aid in navigation, and the paint gives more prominence to the tower structure for passing vessels. Navigation and aerial warning lights also give extra avoidance assurance.

Generally, once a suitable site is located for a wind facility, a foundation style is selected that is compatible with the depth of water, seabed sediment, and rock formation at that site location **FIGURE 10-5**.

Three structural styles for grounding a wind tower and setting a foundation from which the rest of the wind structure is supported include:

- Monopile—Consisting of a single column (steel piling) driven into the seabed 32 to 64 feet, depending on the soil and rock formation conditions. This is the most commonly used foundation to date in the more shallow waters.

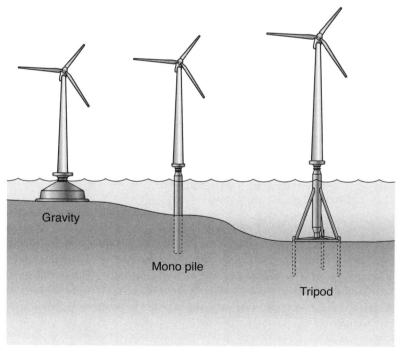

FIGURE 10-5 Three basic types of offshore wind tower foundations.
Adapted from OffshoreWind.net

- Gravity—Foundation constructed from either concrete or steel, which rests on the seabed floor. This system is dependent on gravity to remain in place.
- Tripod—A piling on each foundation corner, driven 32 to 64 feet into the seabed for stability. This design is based on technology used by the oil and gas industry.

Deep-water offshore foundations are still in the conceptual stage with designs being borrowed from the oil and gas industry FIGURE 10-6. Current commercial-scale technology exists with little operational track record. The cost of construction works out to be the biggest challenge facing deep-water offshore siting of wind towers.

Regardless of sea depth and foundation technology, every tower has one thing in common. At the base of the tower above the still-water level sits an enclosed platform providing sheltered access for personnel. This enclosure also provides housing for any remaining backup or support tower and turbine components.

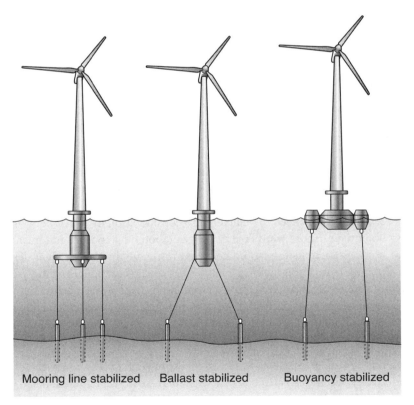

Mooring line stabilized   Ballast stabilized   Buoyancy stabilized

FIGURE 10-6 Prototype floating platform configuration and foundations for deep-water offshore foundations.
Adapted from OffshoreWind.net

A harsh marine environment and the considerable distance from a land base give rise to the necessity for offshore wind towers to be much more rugged and reliable. This distance and the open seas allow designers more generous freedom in tower layout.

Determination of the optimum layout for an offshore wind development concerns itself with many of the same economic factors as that of land-based facilities but with different drivers. Acoustic noise, fixed surface roughness, and tower footprint dominate land-based tower layout, but these are not an issue for offshore projects. The further out from shore, the less acoustic noise and skyline profile are taken into consideration. A tight array of towers over many sea acres will not be seen or heard onshore. Decidedly less surface roughness allows for a much more evenly staggered tower layout than land-based towers; however, seabed conditions are a big consideration.

Research and experimentation have shown that a consistent depth of seawater and homogeneous soil properties have a decisive impact on cost of production and revenue expectations. When water depth and soil properties vary widely, developers must make complex tradeoffs between production, support structure costs, and electrical system costs. Other tradeoffs also must be taken into consideration and put into place, especially in the determination of the optimum array layout.

NOTE

Researchers are also investigating the use of wind resources to produce hydrogen via the hydrolysis of desalinated seawater. This source of energy would then be transported to shore for later use.

A balance must be struck between energy capture losses due to rotor turbulence effect if towers are situated in close proximity versus expansive layouts where cabling costs and O&M cost escalate due to distance between towers. This cost is realized in the undersea electrical collection system of cables running along the seabed floor that connect multiple tower transformers to an offshore substation. Once the undersea cables reach shore they are transferred to an onshore utility grid site for transmission to the land-based power grid.

Several constraints need to be taken into consideration by developers when siting a potential offshore wind facility. To avoid potential conflict and mishaps several siting restrictions for candidate areas are already in place such as designated shipping lanes and pipelines, commercial and recreational fishing areas, low-flying aircraft flight patterns, military operations, and radar systems.

Other potential environmental impacts that may occur and are given major consideration during construction and operation of an offshore wind facility include:

- Marine life—Congregate around and on the tower foundation, which effectively becomes an artificial reef. This increase in fish population and new food supply source may stimulate an increased bird population in the area resulting in possible collisions between birds and turbine rotors.

Developers should consider the electromagnetic fields created by the underwater electric cables running from each tower and to shore. This is a potential threat to the natural benthic communities and could alter the natural seafloor environment. (**Benthic** means relating to anything happening under a body of water or on the sea bottom.) The cabling running from each tower and to shore sets up electromagnetic fields with possible underwater noise and vibration that could affect orientation and navigation ability of surrounding aquatic life.

- Migrating birds—Have the potential for collision with the turbine rotors. Regulators and researchers are questioning if migrating birds need to consume more energy to avoid tower collisions and still maintain their orientation during migration due to having to navigate around offshore tower sites. Tower illumination for shipping lane traffic is also a concern. Regulators question if the illumination could cause navigational disorientation for birds.

- Marine traffic—must be alerted to offshore wind development by use of tower lights, brightly painted platforms, and mapping coordinates.

Offshore wind development currently is dictated by the depth of ocean waters. Most offshore wind farms are located close to their respective shorelines in shallow waters where the technology for tower foundations is already in place but has a limited depth criterion. It is estimated that the commercial deep-water technologies will take another 10–15 years to fully develop.

Currently, the cost of offshore wind-development is almost double that of onshore facilities. These costs are for wind facilities located in shallow water (less than 100 ft. deep). Wind speeds tend to increase with increased distance from shore where 5 MW or larger wind turbines will become the norm. New technologies are being devised and put into place for the infrastructure to support these megawatt turbines.

DEEP OFFSHORE WIND FARMS: CHALLENGES AND OPPORTUNITIES

Stronger winds, less environmental impact, and less impact on real estate combine to give rise to the evolution of wind development toward deep **offshore continental shelf (OCS)** wind farms. An OCS is a stretch of ocean seabed not immediately adjoining or contiguous to a shoreline but further out in deeper water. Currently, OCS wind farms are located in water of 100 ft. or less. A conventional technology used in the oil and gas industry for offshore drilling called

(Continues)

DEEP OFFSHORE WIND FARMS: CHALLENGES AND OPPORTUNITIES (Continued)

jack-up platforms is being revamped for use in the wind industry **FIGURE 10-7**. This technology allows for wind developers to look more seriously at ocean depths greater than 100 ft.

A jack-up platform is actually a floating barge fitted with three or four support legs that are raised (or lowered) by means of a rack and pinion gear system on each leg. The whole structure can be assembled on shore with the tower and turbine in place before being towed to a site location with its legs up. Once on location, the legs are lowered into the seabed and set into place. After being established in place, the jacking system is then used to raise the entire barge above the still-water level leaving an "air gap" underneath the barge. This gives the jack-up system several advantages:

- **Registered marine vessel and totally mobile**—They can be reused on other wind farms during their life cycle. Hydraulics used for installation can be uninstalled and used on other projects. The jack-up can be towed to a safe harbor for serious maintenance alongside a dock.
- **Mass production techniques**—Allows for faster construction, maximizing weather windows.
- **Environmental impact**—In terms of footprint on seafloor, the jack-up system so far appears to have a minimal environmental impact when compared to current foundation technology.

Wind towers mounted on a jack-up platform will stand at a height considerably higher than the current standard monopole, gravity, or tripod style mounted towers. The additional height computes to additional wind energy available for capture.

As wind facilities expand into deeper waters, larger areas of intrusion into the commercial shipping lanes will take place. New lanes will be required with the proper lighting initiated. However, the radar signature of this style system is far more specific. Overall, the jack-up system holds great promise for the wind industry's future deep offshore development needs.

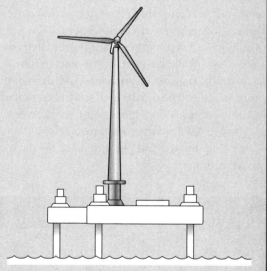

FIGURE 10-7 Three-legged jack-up wind tower.

CHAPTER SUMMARY

Many people in the wind industry believe that offshore wind development is the future for wind energy. Higher energy yields per dollar invested with less impact on the immediate surroundings gives credence to the idea of expanding wind energy production offshore.

To date, offshore wind development has been dictated by the depth and harshness of the ocean floor. With new technologies, the industry is looking further out to sea where the wind energy is even greater. However, as the distance from land increases so do the costs of construction and O&M. To balance the capture of wind power and reach economic viability, it is generally believed that turbines of 5 megawatts or more will become the norm.

KEY CONCEPTS AND TERMS

Benthic

Continental shelf

Offshore

Offshore continental shelf (OCS)

Topographic effect

CHAPTER ASSESSMENT: DESIGN OF OFFSHORE WIND FARMS

1. Which of the following are reasons the United States has not pursued offshore development as diligently as European countries?
   - ❏ A. The United States has more open land readily available for wind development.
   - ❏ B. Technology designed for the European offshore environment needs to be modified and adapted to the harsher US conditions.
   - ❏ C. Cost of construction for land-based facilities is much less than offshore developments.
   - ❏ D. The US coastline is much deeper and more severe than Europe.
   - ❏ E. All of the above.

2. Which of the following explains why offshore wind development versus land-based facilities holds special US interest?
   - ❏ A. More robust environment, allowing higher energy yield for the dollar invested.
   - ❏ B. Much of the offshore potential wind energy is near major populations.
   - ❏ C. Much larger turbines can be considered, as offshore installations do not have the same transportation constraints as land-based turbines.
   - ❏ D. Less obtrusive installation means offshore towers with their apparent size and noise are greatly diminished by distance.
   - ❏ E. All of the above

3. The best US wind potential and also water depths that gradually deepen with distance from their respective shores have been identified as which of the following? (Select two.)
   - ❑ **A.** New England coastline
   - ❑ **B.** Mid-Atlantic coastline
   - ❑ **C.** Midwestern prairie
   - ❑ **D.** Southwest United States

4. Offshore wind towers are interchangeable with land-based towers.
   - ❑ **A.** True
   - ❑ **B.** False

5. Three basic offshore wind tower foundations are: (Select three.)
   - ❑ **A.** monopile
   - ❑ **B.** platform
   - ❑ **C.** gravity
   - ❑ **D.** tripod

6. Deep-water offshore foundations are still in the conceptual stage with designs borrowed from the _____ and _____ industry.

7. Regardless of sea depth and foundation technology, every tower has what one thing in common at its base above the still-water level?
   - ❑ **A.** Mooring dock
   - ❑ **B.** Enclosed platform
   - ❑ **C.** Barges
   - ❑ **D.** Pilings

8. The _____ works out to be the biggest challenge facing deep-water offshore siting of wind towers.

9. Determination of the optimum layout for an offshore wind development concerns itself with many of the same economic factors as that of land-based facilities but with different drivers.
   - ❑ **A.** True
   - ❑ **B.** False

10. Currently, the cost of offshore wind development located in shallow water of less than 100 feet deep is almost _____ that of onshore facilities.
   - ❑ **A.** three times
   - ❑ **B.** 75 percent
   - ❑ **C.** double
   - ❑ **D.** even with

11. Which siting restrictions already in place must be taken into consideration by a developer when siting a potential offshore wind facility to avoid potential conflict and mishaps?
    - ❑ **A.** Designated shipping lanes and pipelines
    - ❑ **B.** Commercial and recreational fishing areas
    - ❑ **C.** Low-flying aircraft flight patterns
    - ❑ **D.** Military operations and radar systems
    - ❑ **E.** All of the above

12. Marine life will congregate around and on the tower foundation, which effectively becomes an artificial reef.
    - ❑ **A.** True
    - ❑ **B.** False

13. A jack-up platform is actually a floating _____ fitted with support legs that are raised (or lowered) by means of a rack and pinion gear system on each leg.

Answer Key

CHAPTER 1 Wind Technology and Design Overview

1. B 2. C 3. A 4. B 5. B 6. C 7. Reduce costs 8. B
9. C 10. Energy

CHAPTER 2 Wind Turbine Technology and Design Concepts, Part 1

1. B 2. C 3. Electricity produced, wind energy converted
4. Blade tip, velocity 5. D 6. A 7. B 8. A 9. C
10. Aerodynamic, mechanical 11. B 12. C

CHAPTER 3 Wind Turbine Technology and Design Concepts, Part 2

1. Generator 2. C 3. B 4. B 5. A 6. B 7. B 8. C
9. A 10. Inductive, synchronous 11. B 12. 1.5 to 35 kV

CHAPTER 4 Design Factors Affecting Weight and Costs

1. C 2. A, B, and D 3. A 4. B 5. B and D 6. A, B, and D
7. Drivers 8. Aerofoil section 9. B 10. A, C, and D
11. A, B, and D 12. B 13. Weibell 14. D 15. A, C, and D

CHAPTER 5 Determining Wind Turbine Weight and Costs

1. A 2. C 3. D 4. A and C 5. A 6. A, B, and D 7. Thrust
8. E 9. Spar 10. C 11. E 12. A, B, and D 13. Spinner
14. A 15. A, C, and D

CHAPTER 6   Weight and Cost of Different Turbine Concepts

1. E **2.** A and B **3.** 45, 35, and 20 **4.** A **5.** C and D **6.** E
7. B **8.** Speed and torque **9.** A, B, and D **10.** B
11. Direct-drive **12.** C **13.** Cross arms **14.** A

CHAPTER 7   Wind Turbine Siting, System Design, and Integration

1. Siting **2.** E **3.** A, B, and C **4.** Midwest and Great Plains
5. A, B, and D **6.** A **7.** B and C **8.** A, C, D, and E **9.** F
10. Array, clustering **11.** B **12.** A **13.** B **14.** 200 **15.** B

CHAPTER 8   Environmental Impact of Wind Systems

1. 10, 15 **2.** A **3.** E **4.** Carcasses **5.** E **6.** B **7.** B, C, and D
8. Motion smear **9.** A **10.** Flicker **11.** B and D **12.** A
13. B **14.** B **15.** E

CHAPTER 9   Economic Prospects of Large Turbines

1. C **2.** A **3.** Fossil fuel, carbon **4.** A and C **5.** B **6.** E
7. B and C **8.** A **9.** C **10.** B **11.** 3, 10 **12.** A
13. Intermittent and cyclic **14.** A, C, and D **15.** A, B, and C

CHAPTER 10   Design of Offshore Wind Farms

1. E **2.** E **3.** A and B **4.** B **5.** A, C, and D **6.** Oil, gas
7. B **8.** Cost of construction **9.** A **10.** C **11.** E **12.** A
13. Barge

GLOSSARY

Ambient At or within a short distance; elements near each other.

Anemometer A device that provides mechanical measurement of wind speed and direction.

Avian Pertaining to or having characteristics of birds.

Barotrauma Rapid pressure change, causing severe, even deadly, internal organ damage.

Bending moment The rotational force of each rotor blade.

Benthic Relating to anything happening under a body of water or sea bottom.

Capital An asset used for the production of further assets.

Coefficient of power (C_p) A number representing a quantity of comparison.

Continental shelf A legal definition denoting the stretch of ocean seabed immediately adjoining and contiguous to the shores of its adjacent country.

Coriolis force The deflection of a force when viewed from a rotating reference. For example, a clockwise rotation will have a deflection to the left of the object in motion.

Cut-in speed Minimum wind speed at which a stalled/stopped turbine begins to rotate its blades and generate usable power, typically between 7 and 10 mph.

Damping Reducing the resonant frequencies of a component or system.

Energy The capacity for work. Energy can be converted into different forms, but the total amount of energy remains the same.

Energy cost The amount of combustible fuel (fossil, nuclear, gas, etc.) required to support the production of a service or product.

Ex-works External components used by the wind turbine but not part of it such as the tower structure, electrical equipment on the ground, and overhead power lines.

Factor A statistical variable whose value is independent of the load and weight formula but relevant to the calculated decision being made.

Flora and fauna Plant life and animal life in a particular region.

Flux The lines of force surrounding a magnetic field, direction of action (i.e., direction perpendicular to the magnetic field and direction of motion).

Footprint Boundary or actual surface area taken up by a structure.

Generator step-up (GSU) A step-up transformer, so called because it is used to step up (i.e., increase) the voltage from the towerhead generator.

Grid A common term referring to an electricity transmission and distribution system; also known also as a "power grid" or "utility grid").

Horizontal axis The axis parallel to the ground in a wind tower, also represented by a reference line through the turbine gears.

Hub The center of a wind turbine rotor that holds the blades in place and attaches to the drive shaft.

Inertia The tendency of a body to maintain its state, either at rest or in a uniform motion, until acted upon by another force.

Integral ($\int$) One of two main operations in calculus denoting an interval of numbers containing all points between two given set points.

Leading edge The edge of a rotor blade that faces toward the direction of rotation (comes in contact with the wind first).

Load   Something physical or electrical that absorbs forces (torque, thrust, moment, vibration, resonant frequencies).

Load path   The path taken by off loading the forces from one component to another until such time as the path route reaches ground level.

Moment   A rotational force around an axis.

Nacelle   Housing for the components of a wind turbine at the towerhead to include generator system, braking, drive train, gears, shaft, and instrumentation. It also acts as the support structure and part of the tower load-path.

Offshore   A designated location away from land or a shoreline in a body of water.

Offshore continental shelf (OCS)   A stretch of ocean seabed not immediately adjoining or contiguous to a shoreline but farther out in deeper water, having, however, a raised seabed, allowing for a shelf of underwater seabed to rise above its surrounding sea floor.

Phase   An electrical state of excitement identical to an adjacent state separated only by a time period in the recurring sequence.

Point of common coupling (PCC)   An Instrumentation and Control (I/C) process at the substation where the wind energy seller (wind farm) and a buyer (utility company) come together to determine whether there is compatibility. If not, the substation will disconnect (main circuit breaker) and can shut the whole wind farm off from the power grid.

Policies   Plan of action (guidelines) adopted by a specific group for actively achieving a set of goals.

Power   Energy that is capable of or available for doing work.

Raptor   A bird of prey such as the eagle, members of the hawk family, and vulture species.

Rated power   The generated power output of a wind turbine in kilowatts.

Repowering   The process whereby an existing wind facility is upgraded with new technology, allowing for higher megawatt production.

Regulations   Prescribed standards of measurement or values that must be met, thereby bringing uniformity to a system (often referred to as an ordinance).

Resonant frequency   Natural inherent low-level vibration produced when a system is in its operational state (the excitable state of a stable component).

Revolutions per minute (rpm)   The number of times a shaft completes a full revolution in one minute.

Rotor   Hub and blade assembly.

Scaling   Size, or change in size, as in reference to scaling models reflecting various sizes of wind turbines.

Shaft   The rotating part in the center of a wind generator or motor that transfers power.

Siting   The locating of a wind farm development.

Slip   Variation in generator rotor revolutions per minute by means of a slip ring.

Step up   To increase in magnitude; used with reference to voltage.

Synchronous   Occurring at the same time (as something else) in a sequential operation.

Tensile strength   Strength of material expressed as the greatest lengthwise stress it can bear without tearing apart.

Tip-speed ratio   The ratio between the rotational speed of the tip of the rotor blade and the actual velocity of the wind.

Topographic effect   Surface configuration of land and the relationship among its man-made and natural features.

Torque   The amount of force it takes to cause an object to rotate. A vector unit where one pound of force acting at a perpendicular distance of one foot from a pivot point is referred to as "foot-pound, commonly abbreviated "ft.-lb."

Tower   Structure used to support the towerhead. People often refer to a tower as a "wind turbine."

Towerhead   A term for the top of the tower encompassing the nacelle and its components, and the rotor assembly. Often referred to as the wind turbine.

Trailing edge   Edge of a blade that faces away from the direction of rotation.

Turbine   The complete assembly unit of rotor blades, hub, drive train, generator, mechanical gears, and instrumentation located in the nacelle. Often referred to as the towerhead of the wind tower.

Turnkey   Describes a facility that is designed, supplied, built, fully installed, and turned over to the developer in production-ready condition, with nothing left to consider or needing attention.

Upwind   The rotor assembly faces the wind in front of the tower as compared to downwind whereby the rotor assembly sits behind the tower in reference to wind direction.

Value engineering   Achieving or surpassing the design criteria specified but for less money.

Vector   A geometric entity specified by two quantities, usually of magnitude and length.

Wind farm   Any number of wind towers within close proximity of one another connecting to a common electrical transfer or substation.

Wind shear   Dramatic shift in wind speed and wind direction over a relatively short distance.

GLOSSARY OF SYMBOLS

C_P Power coefficient (efficiency of power conversion)

C_{Pmax} Maximum power coefficient

d Discounted rate

D Diameter of the rotor assembly established by the circumference of the swept area

E_y Annual energy output of the wind turbine (joules)

P_R Maximum rated power

t Relative profile thickness

V Wind speed

V_d Design wind speed

V_m Annual wind speed at hub height

V_R Rated wind speed

V_T Rotor tip speed

X_d Ratio of design and mean wind speed

X_R Ratio of rated and mean wind speed

Ω Rotational speed of the wind turbine

ρ_a Density of the air (kg/m³)

λ Tip-speed ratio

λ_d Design tip-speed ratio

π Pi (3.14159)

References

10 wind turbines that push the limits of design. (n.d.). *Popular Mechanics.* Retrieved September 24, 2010, from www.popularmechanics.com/science /energy/solar-wind/4324331

Alberts, D.J. (2006). Addressing wind turbine noise. Retrieved December 29, 2010, from http://www.maine.gov

Arnett, E., Huso, M., Schirmacher, M.R., & Hayes, J. P. (2010). Altering turbine speed reduces bat mortality at wind-energy facilities. Retrieved April 26, 2012, from http://www.batsandwind.org

Assessing impacts of wind-energy development on nocturnally active birds and bats. (2007). *Journal of Wildlife Management 71* (8):2449–2486; 2007. Retrieved April 26, 2012, from http://www.batsandwind.org

Blaabjerg, F., & Chen, Z. (2006). *Power electronics for modern wind turbines.* San Rafael, CA: Morgan & Claypool publishers.

Charles F. Brush, Poul la Cour. (n.d.). The Danish Wind Industry Association. Retrieved September 11, 2010, from http://guidedtour.windpower.org/en /pictures/brush.htm

Charles F. Brush. (n.d.). Green Energy Ohio. Retrieved September 11, 2010, from http://www.greenenergyohio.org /page.cfm?pageId=341

Ezawa, B. (2005). Wind–diesel systems in developing countries. Retrieved November 26, 2010, from http://www .wwindea.org/technology/ch05/ estructura-en.htm

Finersh, L., Hand, M., & Laxson, A. (2006). *Wind turbine design cost and scaling model.* Technical Report NREL/TP-500-40566. Retrieved November 1, 2010, from www.nrel .gov/wind/pdfs/40566.pdf

Gray, T. (2010, July 7). "Transpor[t] ation Problems Challenge the Wind Industry." Renewable Energy World. Retrieved January 7, 2011, from http://www.renewableenergyworld .com/rea/news/article/2010/07/wind-transport-needs-challenge-industry

Harnessing the wind. (1941, September 8). Time. Retrieved 11 September, 2010, from http://www.time.com/time /magazine/article/0,9171,849476,00 .html#ixzz0zKZlCvGq

Harrison, R., Hau, E., & Snel, H. (2000). *Large wind turbines design and economics.* West Sussex, England: John Wiley & Sons Ltd.

History of Electric Co-ops. (n.d.). National Rural Electric Cooperative Administration. Retrieved April 26, 2012, from http://www.nreca.org/

Impacting weather, environmental engineers detect turbines' turbulence effects. (2005). Institute of Electrical and Electronics Engineers, Inc., and the American Society of Civil Engineers.

Retrieved October 4, 2010, from www
.sciencedaily.com/videos/2005
/1012-wind_farms_impacting_
weather.htm

Jonkman, J.M. (2003). *Modeling of the UAE
wind turbine for refinement of FAST_
AD,* NREL/TP-500-34755. Retrieved
October 30, 2010, from www.nrel.gov
/docs/fy04osti/34755.pdf

Kubiszewski, I., & Cleveland, C. (2007).
*Energy return on investment (EROI)
for wind energy.* Retrieved January 12,
2010, from http://www.eoearth.org
/article/Energy_return_on_investment_
(EROI)_for_wind_energy

Larson, K. (2008). Making wind more
efficient? Feature article "Energy
Efficiency, Wind Power." Retrieved
November 20, 2010, from www
.renewableenergyfocus.com/view/3271
/making-wind-more-efficient-/.

Meilak, J. (2005). Deep offshore wind
farms – Challenges and opportunities.
Retrieved February 8, 2012, from
http://www.wwindea.org/technology
/ch01/estructura-en.htm

National Renewable Energy Laboratory,
The. (2010). *Eastern wind integration
and transmission study: executive
summary and project overview.*
Retrieved January 8, 2010, from
http://www1.eere.energy.gov
/windandhydro/pdfs/47086.pdf

Offshore wind energy. (n.d.). The OCS
Alternative Energy and Alternate Use
Programmatic EIS Information Center.
Retrieved January 30, 2011, from
http://ocsenergy.anl.gov/guide/wind
/index.cfm

Offshore wind technology (n.d.). US
Department of Energy, Energy
Efficiency and Renewable Energy
(EERE). Retrieved January 29, 2011,
from http://www1.eere.energy.gov
/windandhydro/offshore_wind.html

Offshore wind turbine technology—
Current and future prototypes. (2009).
Retrieved January 30, 2011, from

http://offshorewind.net/Other_Pages
/Turbine-Foundations.html

Plantier, K., & Smith, K. M. (2009).
*Electromechanical principles of wind
turbines for wind energy technicians, 1st
ed.* Waco, TX: TSTC Publishing.

Ragheb, M. (2009). Fatigue loading in
wind turbines. Retrieved on October
18, 2010, from https://netfiles.uiuc
.edu/mragheb/www/NPRE%20475%20
Wind%20Power%20Systems
/Fatigue%20Loading%20in%20
Wind%20Turbines.pdf

Ragheb, M. (2009). Historical wind
machines. Retrieved September 11,
2010, from https://netfiles.uiuc.edu
/mragheb/www/NPRE%20475%20
Wind%20Power%20Systems/
Historical%20Wind%20Generators%20
Machines.pdf

Stallcup, J. G. (2009). *Generator,
Transformer, Motor and Compressor,
2008 Edition.* Sudbury, MA: Jones and
Bartlett Publishers.

Tangler, J. L. (2000). The evolution of rotor
and blade design. National Renewable
Energy Laboratory (NREL). 1617 Cole
Boulevard, Boulder, CO. Retrieved
October 23, 2010, from *www.nrel.gov
/docs/fy00osti/28410.pdf*

Tracton, S. (2009). Can wind farms
change the weather? The Washington
Post. Retrieved October 4, 2010, from
http://voices.washingtonpost.com
/capitalweathergang/2009/07/can_
wind_farms_change_the_weat.html

Transmission and Grid Integration NREL/
FS-550-46187 (2009). Retrieved
January 8, 2010, from http://www1.eere
.energy.gov/windandhydro/pdfs
/46187.pdf

Trout, C. M. (2010). *Essentials of Electric
Motors and Controls.* Sudbury, MA:
Jones and Bartlett Publishers.

US Department of Energy. (2008). 20%
Wind Energy by 2030 – Increasing
wind energy's contribution to US
electricity supply. Retrieved

December 2, 2010, from http://www
.nrel.gov/docs/fy08osti/41869.pdf

US Department of Energy. (2008). 20%
Wind energy by 2030, chapter 5: Wind
power siting and environmental effects,
summary slides. Retrieved December
15, 2010, from http://www1.eere.energy
.gov/windandhydro/pdfs/20percent_
summary_chap5.pdf

US Department of Energy, Federal Wind
Siting Information Center. (2010).
Retrieved December 2, 2010, from
http://www1.eere.energy.gov
/windandhydro/federalwindsiting/
policies_regulations.html

US Department of Energy, Federal
Wind Siting Information Center.
(2010). Retrieved December 4, 2010,
from http://www1.eere.energy.gov
/windandhydro/federalwindsiting/pdfs
/windmill_policy_letter_012907.pdf

US Department of Energy, Federal
Wind Siting Information Center.
(2010). Retrieved December 4, 2010,
from http://www1.eere.energy.gov
/windandhydro/federalwindsiting/pdfs
/windmill_policy_letter_071006.pdf

US Department of Energy, Federal
Wind Siting Information Center.
(2010). Retrieved December 4, 2010,
from http://www1.eere.energy.gov
/windandhydro/federalwindsiting/pdfs
/windmill_policy_letter_032106.pdf

US Department of Energy, Federal Wind
Siting Information Center. (2010).
Retrieved December 7, 2010, from
http://www.fs.fed.us/recreation
/permits/documents/federal_register_
wind.pdf

US Department of Energy, Federal
Wind Siting Information Center.
(2010). Retrieved December 8, 2010,
from http://www1.eere.energy.gov
/windandhydro/federalwindsiting/pdfs
/ntia_to_irac.pdf

US Department of Energy. *(n.d.).*
Renewable systems interconnection.
Retrieved January 7, 2011, from
http://www1.eere.energy.gov/
windandhydro/renewable_systems.html

US Department of Energy, Wind & Power
Program. Retrieved December 15, 2010,
from http://www1.eere.energy.gov
/windandhydro/impacts_siting.html

US Department of Energy. *Wind Powering
America.* Retrieved December 2, 2010,
from http://www.windpoweringamerica
.gov/siting.asp

Vibration Testing: NREL, National
Laboratory of the U.S. Department of
Energy. Retrieved September 12, 2010,
from http://www.nrel.gov/docs/legosti
/fy96/21218.pdf

Wahl, D., and Giguere, P. Ice shedding
and ice throw – risk and mitigation.
Retrieved December 31, 2010, from
http://www.gepower.com/prod_serv
/products/tech_docs/en/downloads
/ger4262.pdf

Wind Energy – Technology and Planning
(n.d.). World Wind Energy Association
(WWEA). Retrieved January 29, 2011,
from http://www.wwindea.org
/technology/ch01/estructura-en.htm

Wind energy: The facts. (n.d.). Retrieved
November 26, 2010 from http://www
.wind-energy-the-facts.org/ro
/part-i-technology/chapter-3-
wind-turbine-technology/current-
developments/alternative-drive-train-
configurations.html

Wind turbine interactions with birds,
bats, and their habitats: A summary of
research results and priority questions.
(Spring 2010). National Wind
Coordinating Collaborative. Retrieved
December 23, 2010, from www
.nationalwind.org

World wind energy report. (2009).
American Wind Energy Association.
(2009). Retrieved on October 18, 2010,
from www.wwindea.org/home/images
/stories/worldwindenergyreport2009_
s.pdf

Index

Figures and tables are indicated by f and t following page numbers.